Feedback Circuits and Op. Amps

TUTORIAL GUIDES IN ELECTRONIC ENGINEERING

This series is aimed at first and second year undergraduate courses. Each text is complete in itself, although linked with others in the series. Where possible, the trend towards a 'systems' approach is acknowledged, but classical fundamental areas of study have not been excluded. Worked examples feature prominently and indicate, where appropriate, a number of approaches to the same problem.

A format providing marginal notes has been adopted to allow the authors to include ideas and material to support the main text. These notes include references to standard mainstream texts and commentary on the applicability of solution methods, aimed particularly at covering points normally found difficult. Graded problems are provided at the end of each chapter, with answers at the end of the book.

1. Transistor Circuit Techniques: discrete and integrated (2nd edition) — G.J. Ritchie
2. Feedback Circuits and Op. Amps (2nd edition) — D.H. Horrocks
3. Pascal for Electronic Engineers (2nd edition) — J. Attikiouzel
4. Computers and Microprocessors: components and systems (2nd edition) — A.C. Downton
5. Telecommunication Principles (2nd edition) — J.J. O'Reilly
6. Digital Logic Techniques: principles and practice (2nd edition) — T.J. Stonham
7. Transducers and Interfacing: principles and techniques — B.R. Bannister and D.G. Whitehead
8. Signals and Systems: models and behaviour — M.L. Meade and C.R. Dillon
9. Basic Electromagnetism and its Applications — A.J. Compton
10. Electromagnetism for Electronic Engineers — R.G. Carter
11. Power Electronics — D.A. Bradley
12. Semiconductor Devices: how they work — J.J. Sparkes
13. Electronic Components and Technology: engineering applications — S.J. Sangwine
14. Optoelectronics — J. Watson
15. Control Engineering — C. Bissell
16. Basic Mathematics for Electronic Engineers: models and applications — J.E. Szymanski
17. Software Engineering — D. Ince

Feedback Circuits and Op. Amps

Second edition

D.H. Horrocks
Senior Lecturer, School of Electrical, Electronic and Systems Engineering, University of Wales College of Cardiff

Chapman and Hall
University and Professional Division

First published in 1983 by
Van Nostrand Reinhold (International) Co. Ltd

Reprinted 1984, 1985

Second edition published in 1990 by
Chapman and Hall Ltd,
11 New Fetter Lane, London EC4P 4EE

Typeset in Singapore by Colset Pte Ltd

Printed in Hong Kong

ISBN 0 412 34270 7
ISSN 0266 2620

British Library Cataloguing in Publication Data

Horrocks, D.H. (David H.)
Feedback circuits and op. amps. – 2nd ed.
1. Electronic equipment. Feedback circuits
I. Title
621.381535

ISBN 0 412 34270 7

Preface

Feedback circuits in general, and op. amp. applications which embody feedback principles in particular, play a central role in modern electronic engineering. This importance is reflected in the undergraduate curriculum where it is common practice for first-year undergraduates to be taught the principles of these subjects. It is right therefore that one of the tutorial guides in electronic engineering be devoted to feedback circuits and op. amps.

Often general feedback circuit principles are taught before passing on to op. amps., and the order of the chapters reflects this. It is equally valid to teach op. amps. first. A feature of the guide is that it has been written to allow this approach to be followed, by deferring the study of Chapters 2, 4 and 5 until the end.

A second feature of the guide is the treatment of loading effects in feedback circuits contained in Chapter 5. Loading effects are significant in many feedback circuits and yet they are not dealt with fully in many texts.

Prerequisite knowledge for a successful use of the guide has been kept to a minimum. A knowledge of elementary circuit theory is assumed, and an understanding of basic transistor circuits would be useful for some of the feedback circuit examples.

I am grateful to series editor Professor G.G. Bloodworth for many useful discussions and suggestions. I am also grateful for the expert typing of Mrs. B. Richards and for help in preparing solutions to the problems given by Mr. R. Hor. Finally, I thank my wife, Chris, and son, Jim, for their forbearance.

With this second edition the opportunity has been taken to enhance the first edition in several ways. Chapter 9 has been added to extend significantly the range of op. amp. circuit applications covered by the tutorial guide. Further applications of feedback and op. amps. can be explored through the bibliography which has been included. The text makes some use of phasors and the j-notation with which some readers may not be familiar. To help such readers, an appendix introducing these topics is provided.

Contents

Preface v

1 Introduction 1

2 General Properties of Feedback Amplifiers 4
Basic definitions and equations 4
Positive and negative feedback 7
Reduced sensitivity to gain variations 11
Reduction of noise and distortion 13
Multiple-loop feedback 15

3 Amplifiers Without Feedback 21
Amplifier models 21
Connection of signal-source, amplifier and load 26
Frequency response effects 29
Differential amplifiers 39

4 Feedback Amplifier Circuits 47
The four feedback circuit configurations 47
Effect of feedback on input and output impedance 49
Shunt-voltage feedback circuit example 56
Series-voltage feedback circuit example 59
Series-current feedback circuit example 61
Shunt-current feedback circuit example 64

5 More About Feedback Amplifiers 69
Loading effects between forward amplifier and feedback block 69
Effect on feedback amplifier of source and load impedances 82
Frequency response of feedback amplifiers 88
Instability 90

6 The Op. Amp. — Basic Ideas and Circuits 97
What is an operational amplifier? 97
Inverting voltage amplifier 100
Circuits based on the inverting voltage amplifier 105
Non-inverting voltage amplifier 108
Power supply connections to the op. amp. 111

7 Op. Amp. Non-idealities 115
The importance of op. amp. non-idealities 115
Offset voltages 115
Bias currents and offset currents 120
Op. amp. frequency response 124
Slew rate and full-power bandwidth 126

8 Selected Op. Amp. Applications 131
Precision difference and instrumentation amplifiers 131
Analogue computation 135
Wien bridge oscillator 139
The inverse function principle 142
Triangle wave/square wave generator 145

9 Further Op. Amp. Applications 149
Circuits derived from the precision difference amplifier 149
A.C. amplifiers 156
Active filters 161
Precision rectifier circuits 170

Appendix: Steady-State Network Analysis using Phasors and Complex Variables 176

Bibliography 184

Answers to Numerical Problems 185

Index 187

Introduction

Many examples of feedback can be found in everyday life. One example is the temperature regulation of a heated room, shown schematically in Fig. 1.1. The heater supplies heat to the room and the temperature of the room, T_M, is sensed. This is fed back and compared with the desired temperature, T_D. The difference, $T_E = T_D - T_M$ (called the *error signal*), is passed to the regulator. The regulator uses the error signal to control the heater to maintain the room temperature close to the desired temperature. This automatic control technique is an example of *negative feedback*, so called because, in generating the error signal, the function of the comparator is to *subtract* the measured temperature from the desired temperature.

The subject of this tutorial guide is the application of feedback principles to electronic amplifying circuits containing active devices such as transistors. The use of integrated circuit transistor amplifiers, called *op. amps.* (an abbreviation of *operational amplifiers*), is particularly important and is considered in some detail.

The elements of a negative feedback amplifier system are shown in Fig. 1.2. The amplifying circuit uses active devices, normally transistors, to increase the magnitude of the electronic signal applied to its input. The input signal to the amplifier comes from a comparator which subtracts from the external input signal a fraction β of the amplifier output signal. If the amplifier has high gain it requires only a small signal at its input. Therefore the external input signal is approximately equal to the fraction β of the output signal. It follows that the output signal must be approximately equal to the input signal divided by β. This technique gives a system with a stable overall gain if the components in the feedback circuit have stable values. This is relatively easy to achieve using passive components such as resistors. The amplifying circuit itself does not need to have stable gain (but the gain must be high). This is of importance because the stable and predictable gain of systems is highly desirable for most electronic systems, but active devices so far invented do

A very significant result.

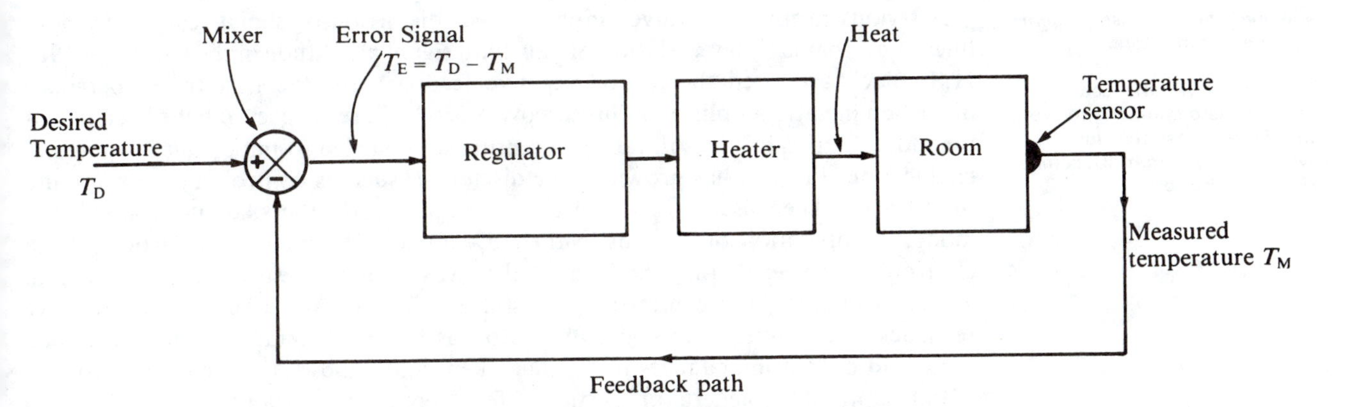

Fig. 1.1 Temperature regulation of a heated room.

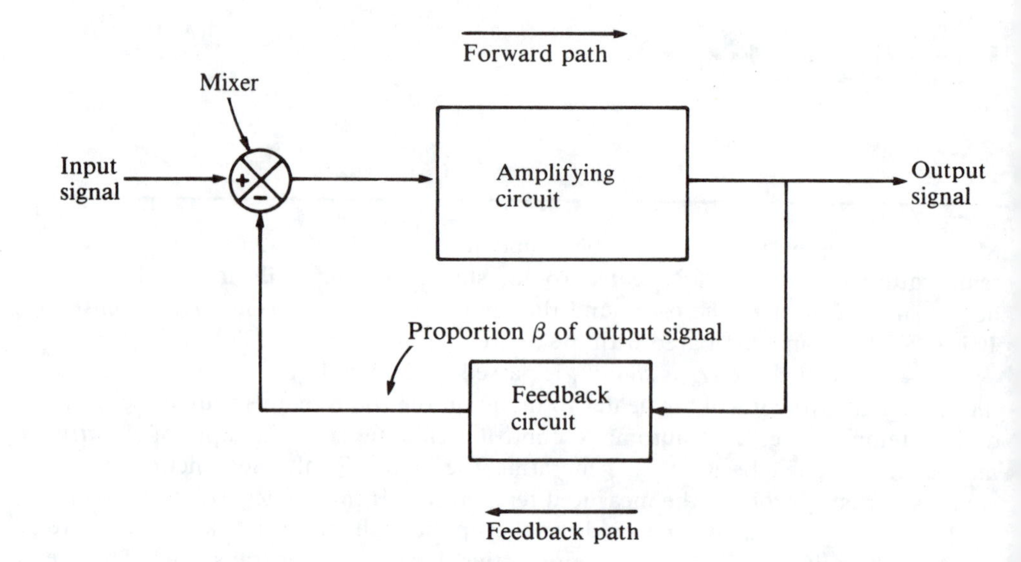

Fig. 1.2 Operating principle of a negative feedback amplifier system.

not have stable characteristics. Fortunately the requirement for the feedback system that the amplifier should have high gain is readily and cheaply achieved nowadays with transistors, particularly in integrated circuit form.

It should be noted that the stability of gain provided by negative feedback is not merely stability with the passage of time. Negative feedback also reduces the variations of gain with environmental conditions such as temperature, with manufacturing component spreads, and with the frequency and amplitude of the signals.

From Fig. 1.2 it is apparent that if a fraction β of the output signal is *added* to the input signal the overall gain is increased. This is called *positive feedback*.

During the early development of thermionic valve amplifiers positive feedback was used to increase the gain because valves were expensive and had low performance.

Non-linearities are also present in modern transistors.

Black conceived the idea 'in a flash' while commuting to work on a ferry across the Hudson river, *IEEE Spectrum*, December 1977, pp. 56–60.

A serious problem of valve amplifiers was distortion of signals caused by non-linearities, that is to say variation of gain with signal amplitude in the valves. In 1927 H.S. Black proposed the use of negative feedback for the first time, to reduce distortion in valve amplifiers. This discovery is seen in retrospect to have been one of the most significant ideas in the 20th century and has led to important advances in engineering. From it has grown whole disciplines such as control engineering, and the idea has been used by biologists, economists and others, to understand and model the operation of systems in their disciplines. However, it is particularly in electronic engineering that the idea is all-pervasive as transistors can provide high gain very cheaply, but cannot provide stable gain. This book shows that negative feedback not only stabilizes gain but also has other advantages such as enabling input and output impedances to be controlled, and bandwidth to be extended.

In Chapter 2 the general properties of feedback amplifiers and some of the main advantages of negative feedback are analysed. An introduction to the analysis of multi-loop systems is also included.

Before applying these ideas to amplifier circuits in Chapters 4 and 5, a description is given in Chapter 3 of the relevant properties of amplifier circuits without feedback. This includes calculation of voltage and current gains, multiple stages, input and output impedances and frequency response.

In Chapter 4 the four basic ways to apply negative feedback to an amplifier are explained and some examples are analysed. The simple treatment given in this chapter covers cases where loading effects can be neglected.

The methods of taking these effects into account are explained in Chapter 5, which also deals with the frequency response of feedback amplifiers. This material is more advanced and may be omitted from a first reading.

The remainder of the book deals with operational amplifier circuits. The name arises from the use of this type of amplifier in the early development of analogue computers to perform various mathematical *operations* on analogue voltages. The early op. amps. were constructed from thermionic valves, but since they were expensive their application was limited. The development of the silicon integrated circuit in the nineteen sixties enabled cheap high-gain op. amps. to be mass-produced. Nowadays they are widely used as components with negative feedback to perform many functions in analogue instrumentation, communications, etc.

The subject of analogue computers is briefly covered in Chapter 8.

Chapter 6 explains the characteristics of op. amps. and their use in voltage amplifiers. Use is made of the concept of the ideal op. amp.

Although op. amp. characteristics do approach this ideal they have some limitations which can be significant. In Chapter 7 the nature of the more important non-idealities are described and methods for allowing for them are explained.

In Chapters 8 and 9 some more op. amp. circuits are described which have a range of useful applications and illustrate some further principles.

The chapter sequence adopted here follows one of the logical ways of developing and studying the subject at an introductory level. However, an equally logical and valid method of approach is to study op. amp. circuits before studying discrete transistor circuits. A reader following the latter method of study should defer a reading of Chapters 2, 4 and 5 until the end.

2 General Properties of Feedback Amplifiers

Objectives

- ☐ To calculate the gain of an amplifier after feedback has been applied.
- ☐ To distinguish between positive and negative feedback.
- ☐ To explain loop-gain and feedback factor.
- ☐ To explain how reduced sensitivity to component variations can be obtained by negative feedback.
- ☐ To describe how to use negative feedback to reduce noise and distortion.
- ☐ To simplify and analyse simple multiple-loop systems.

Basic Definitions and Equations

Feedback is a general idea.

Fig. 2.1 shows a feedback system, with the various signals indicated. In this text the system is assumed to be electrical and the signals marked represent currents or voltages; however, it should be remembered that feedback can be applied to other systems, in which case the signals may represent velocity, cash flow, production output and so on. In the forward path, block A represents an amplifier whose output X_o is equal to the amplifier input signal X_{ia} multiplied by the *forward-path gain* constant A. The gain constant is usually large and can be positive or negative (the latter causing phase reversal of a sinusoidal signal). In general though, as explained later, it can be a complex number which is a function of signal frequency. The output of the amplifier X_o, which is also the output of the feedback system, is sensed by the feedback block. This block provides a feedback signal X_f equal to the

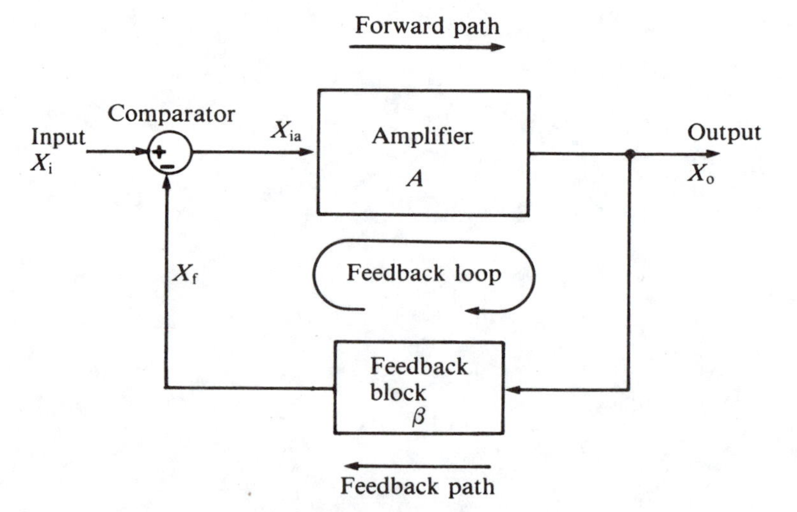

Fig. 2.1 Signals in a feedback system.

system output multiplied by the *feedback fraction* β. The feedback fraction is usually quite small and like the constant A it can be positive, negative or a complex number which is a function of signal frequency. The feedback signal X_f is subtracted from the system input X_i by the *comparator* to give the amplifier input X_{ia}.

Do not confuse β here with the current amplification of a transistor.

In practice an electronic amplifier could be made from a simple amplifying circuit containing a single amplifying device such as a bipolar junction transistor or a field-effect transistor. Or it could be made from a combination of such simple circuits, possibly in integrated circuit form. There is a wide range of possible amplifier circuits and a number of different networks for the feedback block are available. In addition, there is more than one way to compare the feedback signal and the system input signal. These possibilities are explored later in Chapter 4. These details are not required in this chapter as only the general properties of systems with feedback are being considered.

When analysing feedback systems the relationship between input signal X_i and signal output X_o, after feedback has been applied is found first. From this other properties of a feedback system can be discovered.

From Fig. 2.1 it can be seen that the following three equations define the signal inter-relationships.

For the feedback block

$$X_f = \beta X_o \qquad (2.1)$$

For the comparator

$$X_{ia} = X_i - X_f \qquad (2.2)$$

For the amplifier block

$$X_o = A\,X_{ia} \qquad (2.3)$$

To obtain the overall relationship between input X_i and output X_o the internal signals X_f and X_{ia} need to be eliminated from this set of equations. One way is to substitute the first equation into the second to give, for the comparator,

There are three equations in four variables; they can be reduced to one equation in two variables.

$$X_{ia} = X_i - \beta X_o \qquad (2.4)$$

and then to substitute this equation into Equation 2.3 to give:

$$X_o = A(X_i - \beta X_o)$$

Hence

$$X_o(1 + A\beta) = A\,X_i$$

and from this

$$X_o = \frac{A}{1 + A\beta} \cdot X_i$$

The ratio X_o/X_i is called the gain with feedback, A_f. It is also called the *closed-loop gain*, and from the last equation is given by

$$A_f = \frac{X_o}{X_i} = \frac{A}{1 + A\beta} \qquad (2.5)$$

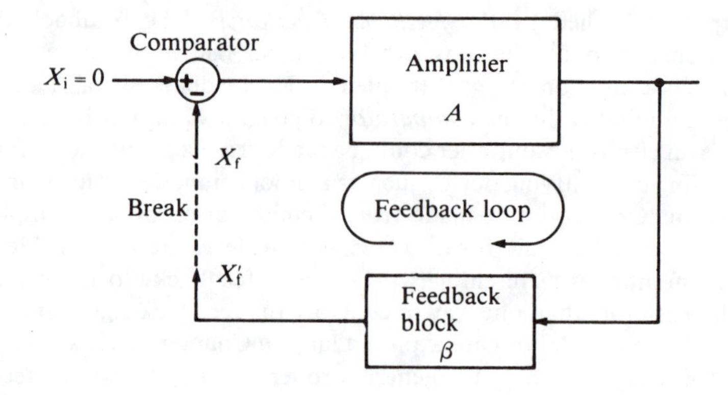

Fig. 2.2 Breaking the feedback loop to obtain loop gain.

This is the *fundamental feedback relationship*. The gain A without feedback is known as the *open-loop gain*.

It can be seen that the effect of feedback is to cause the gain in the forward path to be divided by the factor $1 + A\beta$.

A very important property.

At this point it may be noted that if the magnitude of $A\beta$ is much greater than unity then the closed-loop gain is approximately equal to $1/\beta$, and therefore becomes independent of the properties of the forward amplifier.

Exercise 2.1 Derive the following expressions for the internal signals in terms of the system input;

$$X_f = \frac{A\beta}{1 + A\beta} . X_i \text{ and } X_{ia} = \frac{1}{1 + A\beta} . X_i \tag{2.6}$$

Now consider again the feedback loop in Fig. 2.1 comprising the forward-path amplifier, the feedback block and the comparator. Suppose the system input is not present, $X_i = 0$, and also suppose the feedback loop is temporarily broken at any point, say at the output of the feedback block, as shown in Fig. 2.2. If a hypothetical signal is inserted at point X_f, then the output of the comparator is equal to $-X_f$, (see Equation 2.2 and set $X_i = 0$). This signal passes through the forward amplifier with gain A and emerges at the output with a value $-AX_f$. Then it passes into the feedback block which multiplies the signal by β to give the open-loop feedback signal $X_f' = -A\beta X_f$. Notice that in passing round the loop the hypothetical injected signal, X_f, has been multiplied by the factor $-A\beta$. This factor is called the *loop gain*, thus

This is a *thought experiment*, carried out to understand a concept rather than to do something in practice.

$$\text{Loop-gain} = -A\beta \tag{2.7}$$

Exercise 2.2 Verify that the above argument provides the same formula for loop-gain no matter at what point the loop is broken.

To conclude this introduction we restate the most important result,

The fundamental feedback formula.

$$\text{Closed-loop gain} = \frac{\text{Forward gain}}{1 - \text{Loop gain}}$$

i.e.

$$A_f = \frac{A}{1 + A\beta} \tag{2.8}$$

Positive and Negative Feedback

We are now ready to define the terms *negative feedback* (NFB) and *positive feedback* (PFB). As stated in the introductory chapter, if the input signal to the system is reduced in magnitude when the feedback signal is subtracted from it then the feedback is negative. Looking at Equation 2.6 for the output from the comparator, X_{ia}, in terms of the signal input, X_i, negative feedback occurs when $|1 + A\beta|$ is greater than unity. This fact, when applied to the fundamental feedback relationship (2.8), shows for negative feedback that the magnitude of the gain, with feedback, is less than the forward gain. Converse statements can be made about positive feedback. Therefore the equivalent definitions of NFB and PFB can be written in terms of the magnitudes of the various quantities, as shown in Table 2.1.

Despite the name, negative feedback has very positive benefits!

Table 2.1 Equivalent Definitions of Negative and Positive Feedback

Negative feedback	Positive feedback
$\lvert X_{ia}\rvert < \lvert X_i\rvert$	$\lvert X_{ia}\rvert > \lvert X_i\rvert$
$\lvert 1 + A\beta\rvert > 1$	$\lvert 1 + A\beta\rvert < 1$
$\lvert A_f\rvert < \lvert A\rvert$	$\lvert A_f\rvert > \lvert A\rvert$

The vertical bars | | mean absolute value for real numbers, or modulus for a complex number.

From the formula $(1 + A\beta)$, it can be seen that the presence of negative or positive feedback depends on the particular values of forward gain A and feedback fraction β. A variety of conditions are possible. Consider what happens to the closed-loop gain A_f when, for some particular value of forward-gain A, the feedback fraction β is varied over positive and negative values. The fundamental feedback equation (2.8) is expressed graphically in Fig. 2.3. Although the curve is drawn for A positive, consider five regions of the curve for both positive and negative A.

Region (i) ($A\beta > 0$) (negative feedback). In this region, the loop gain $(-A\beta)$ is negative and so $|1 + A\beta| > 1$, thus showing that the feedback is negative. As β is increased the factor $|1 + A\beta|$ is also increased, the negative feedback strengthened and so the closed-loop gain is reduced. As β decreases towards zero, the closed-loop gain increases in magnitude until at $\beta = 0$, the factor $|1 + A\beta|$ is unity and the fundamental feedback relationship indicates that the closed-loop gain equals the open-loop gain, $A_f = A$. This is to be expected since if $\beta = 0$ it is equivalent to having no feedback at all.

Splitting the curve into smaller parts.

Region (ii) ($-1 < A\beta < 0$) (positive feedback). In this region the loop-gain $(-A\beta)$ is positive and $|1 + A\beta| < 1$, thus showing that the feedback is positive. The fundamental feedback relationship shows that the magnitude of the closed-loop gain is greater than that of the open-loop gain. If β is varied to make $A\beta$ approach -1 then $|1 + A\beta|$ becomes very small and the closed-loop gain increases without limit.

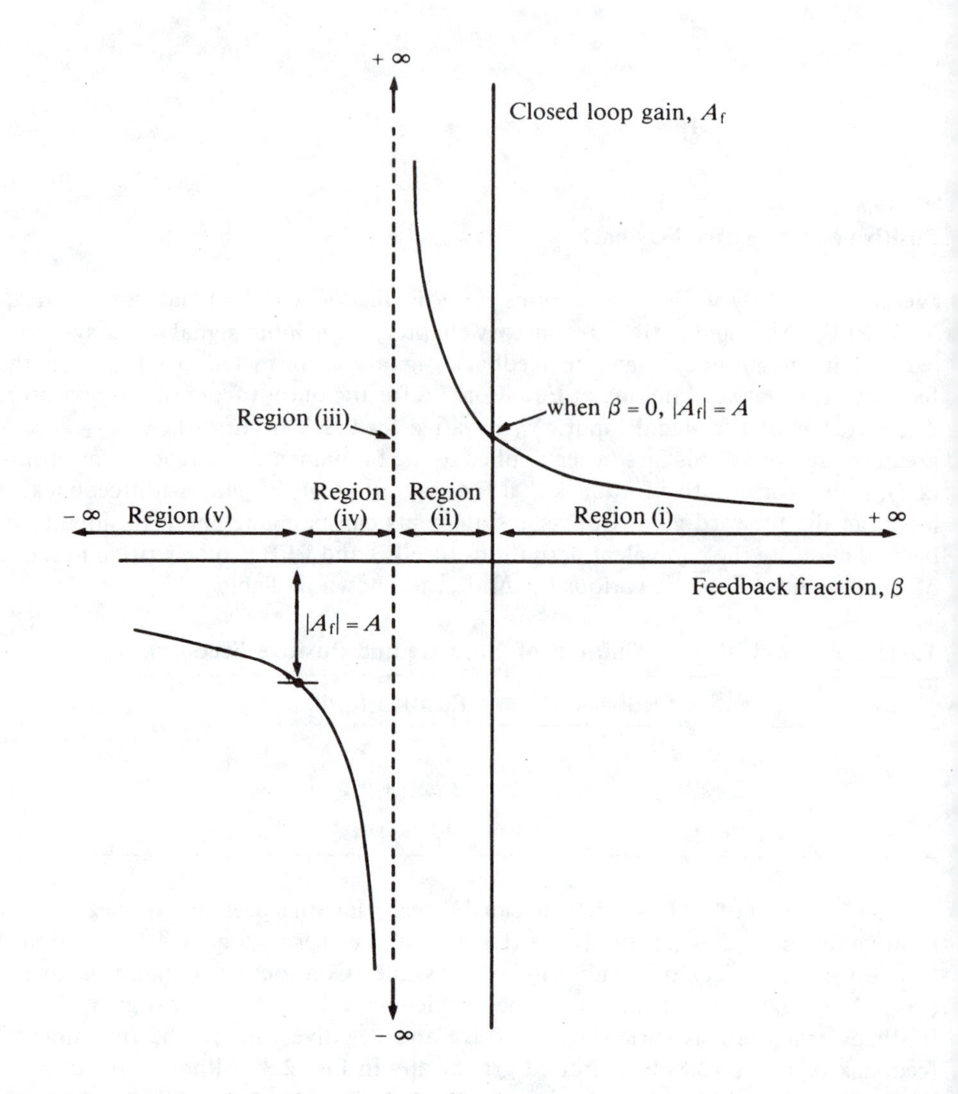

Fig. 2.3 Graph of fundamental feedback equation showing various feedback regions.

We have hit a *pole* of the function for closed-loop gain.

Region (iii) ($A\beta = -1$) (infinite gain, positive feedback). At this point the loop gain equals $+1$ and $|1 + A\beta| = 0$. The fundamental feedback relationship indicates a value of $A_f \rightarrow \infty$. Turning the gain relationship $X_o = A_f.X_i$ around, then $X_i = (A_f)^{-1}.X_o$. Thus $A_f \rightarrow \infty$ shows that a finite output X_o can be sustained for zero input signal. Consider again Fig. 2.2. A loop gain of exactly $+1$ means that $X_f' = X_f$. Thus once the loop is closed any signal in the system goes round the loop for ever with sustained amplitude. In effect the feedback system is not an amplifier of signals but a generator of signals. This is a special case of a feedback amplifier which can be exploited in the design of sinewave oscillators (see Chapter 8).

Region (iv) ($-2 < A\beta < -1$) (positive feedback, usually unstable). Increasing $A\beta$ negatively below -1 further increases the loop-gain beyond $+1$. Thus signals circulate round the feedback loop and usually increase in amplitude in an unstable way without limit until stopped by the inability of the amplifier to handle the large signals thus generated. This behaviour in amplifiers is called *instability* and is undesirable.

Region (v) ($A\beta < -2$) (negative feedback, usually unstable). In this region $(1 + A\beta) < -1$ and so $|1 + A\beta| > 1$ thus showing the feedback to be negative (see Table 2.1). However, as in the previous region (iv) because the loop-gain is greater than $+1$ in this region the system can be expected to be unstable.

Note that from the last two regions the conclusion can be drawn that a positive loop gain, greater than one is expected to result always in an unstable feedback system owing to signals circulating round the feedback loop and growing. However, under special circumstances stable operation is possible (but rarely used in practice). Discussion of this is left until Chapter 5.

In the discussion so far the parameters A and β of the feedback system have been assumed to have positive or negative real values. Signals X_i, X_o and X_f would all be in- or anti-phase if sinusoidal. In practice, the output signal of amplifiers and linear circuits generally are changed in phase as well as in magnitude when compared with the input signal. Both changes are functions of frequency. (The reason is discussed in the next chapter but need not concern us here.) Thus the amplification A_f is now written as the complex variable $\hat{A}_f$ which can be expressed in modulus and angle form or in terms of real and imaginary parts.

An introduction to the use of complex variables is to be found in the appendix.

Nearly all the analysis derived, so far, for feedback systems with real quantities can be used equally well for complex quantities by merely interpreting the various parameters as complex variables and applying the usual rules for manipulation of complex numbers. For example the fundamental feedback equation (2.8) becomes

$$\hat{A}_f = \frac{\hat{A}}{1 + \hat{A}\hat{\beta}} \tag{2.9}$$

and the definitions of negative and positive feedback become

$$|1 + \hat{A}\hat{\beta}| > 1 \text{ for NFB}$$

and

$$|1 + \hat{A}\hat{\beta}| < 1 \text{ for PFB} \tag{2.10}$$

where the vertical bars are now taken to mean modulus of the complex quantity within.

The only part of the previous discussion which does not completely fit the general case of complex quantities is the discussion of the various stability regions of Fig. 2.3. In this figure A and β are assumed to be positive or negative real quantities, with phase shifts of 0° or 180°. The figure does not exist for complex values of $\hat{A}$ and $\hat{\beta}$ and a different approach is required to examine amplifier stability and instability (see Chapter 5).

Worked Example 2.1

An amplifier-block having a nominal gain of $A = 1000$ is to be used in a feedback circuit to provide a closed-loop gain having a magnitude equal to 10. Calculate a suitable value for feedback fraction β. If in practice the amplifier block turns out to have $A = 900\underline{/-30°}$ then calculate the closed-loop gain actually obtained.

Solution. For the first part, a closed-loop gain of either $+10$ or -10 satisfies the requirement for a closed-loop gain magnitude of 10. Substituting values into the fundamental feedback equation (2.8) gives

$$\pm 10 = \frac{1000}{1 + 1000 \times \beta}$$

Solving this gives $\beta = +0.099$ or -0.101. The second of these results in a loop gain of $-A\beta = +101$. This corresponds to region (v) of Fig. 2.3 and because it usually results in instability this value of β is rejected. Hence, $\beta = +0.099$ is chosen, and the closed-loop gain is $+10$.

For the practical case of $A = 900\angle -30^\circ$ then substituting in Equation 2.8

Basic complex variable theory.

$$A_f = \frac{900\angle -30^\circ}{1 + (900\angle -30^\circ)(0.099)} = \frac{900\angle -30^\circ}{1 + 89.1\angle -30^\circ}$$

$$= \frac{900\angle +30^\circ}{1 + (77.16 - j44.55)} = \frac{900\angle -30^\circ}{78.16 - j44.55}$$

$$= \frac{900\angle -30^\circ}{89.96\angle -29.68^\circ}$$

Hence $A_f = 10.004\angle -0.32^\circ$.

This result is still close to the $+10$ value despite the change in magnitude and phase of A. This might have been expected since because $A\beta >> 1$ the closed-loop gain is almost independent of A. In particular, it may be noted that the phase of A_f is very small despite the large phase angle in A because the feedback fraction β has zero phase angle.

Before leaving this section it should be noted that the comparator shown in Fig. 2.1 subtracts the feedback signal from the incoming signal. Some textbooks assume addition of the two signals going into the comparator, as shown in Fig. 2.4. It can be seen from Fig. 2.1 that the negative sign on the comparator can be changed to positive without affecting the general system behaviour, provided X_f and β are relabelled $-X_f$ and $-\beta$, respectively. The existing analysis for Fig. 2.1 can be used for the arrangement in Fig. 2.4 by changing the signs of β and X_f where they occur in the equations. This leads to $1 - A\beta$ instead of $1 + A\beta$ in the basic equations. For consistency the feedback arrangement of Fig. 2.1 is used throughout this text.

Changing the signs in this way saves having to repeat all of the analysis for this case.

Exercise 2.3 Show that for Fig. 2.4, the following hold; loop gain $= (A\beta)$, closed-loop gain, $A_f = A/(1 - A\beta)$.

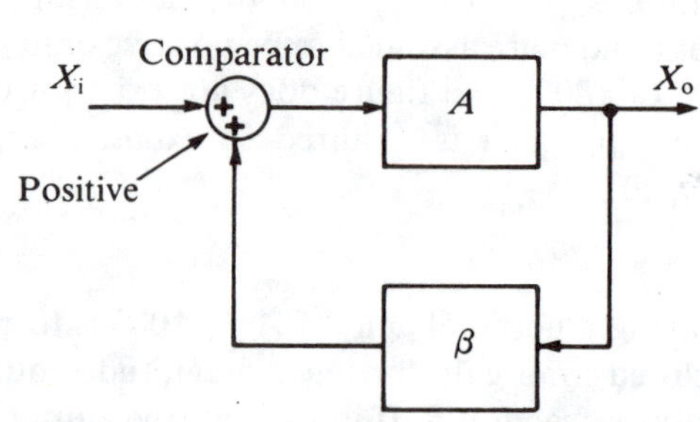

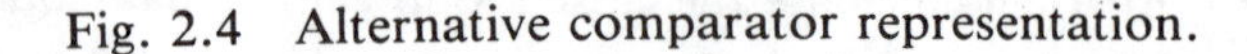

Fig. 2.4 Alternative comparator representation.

Reduced Sensitivity to Gain Variations

In practice the gain of the forward amplifying block A can deviate from the nominal value. This can happen for a number of reasons. For example the parameters of the active and passive components which make up the amplifier can vary owing to temperature and ageing, or can be affected by variations in the supply voltages which provide power to the amplifier. Generally, when a quantity of amplifiers are manufactured no two have exactly the same gain; the gains deviate from the nominal or average value by differing amounts. These production spreads can be quite pronounced because active devices such as bipolar transistors cannot be mass-produced with identical characteristics.

Practical effects.

In the previous worked example it is seen that if the forward amplifier has a gain of $900\underline{/-30^\circ}$ instead of $1000\underline{/0^\circ}$ the resulting closed-loop gain of $10.004\underline{/-0.32^\circ}$ is still very close to the desired value of $10\underline{/0^\circ}$. This is obviously desirable, but is it a general property of negative feedback? To find out, first define some terms. Suppose the forward gain changes by a small amount $\mathrm{d}A$, then the relative change in A is $\mathrm{d}A/A$. Suppose the change $\mathrm{d}A$ causes a small change $\mathrm{d}A_f$ in the closed-loop gain, then the relative change in A_f is $\mathrm{d}A_f/A_f$. To show the effect of feedback consider the relative change in A_f caused by a relative change in A. The quantity relating the two is the *sensitivity parameter*:

Moschytz, G.S. *Linear Integrated Networks: Fundamentals* (Van Nostrand Reinhold, 1974) has a wider and deeper treatment of sensitivity.

$$\frac{\mathrm{d}A_f}{A_f} = S_A^{A_f} \cdot \frac{\mathrm{d}A}{A} \qquad (2.11)$$

where $S_A^{A_f}$ is the sensitivity of A_f with respect to variations in A. Rearranging this equation gives

$$S_A^{A_f} = \frac{\mathrm{d}A_f}{A_f} \Big/ \frac{\mathrm{d}A}{A} = \frac{\mathrm{d}A_f}{A_f} \cdot \frac{A}{\mathrm{d}A} = \frac{A}{A_f} \cdot \frac{\mathrm{d}A_f}{\mathrm{d}A} \qquad (2.12)$$

This provides a convenient way to calculate sensitivity. Substituting for A_f using the fundamental feedback Equation 2.8 gives:

$$S_A^{A_f} = \frac{A}{A/(1 + A\beta)} \frac{\mathrm{d}}{\mathrm{d}A}\left\{\frac{A}{1 + A\beta}\right\}$$

Using the well known rule for differentiating a quotient.

$$S_A^{A_f} = (1 + A\beta)\left\{\frac{(1 + A\beta)(1) - (A)(\beta)}{(1 + A\beta)^2}\right\}$$

That is

$$S_A^{A_f} = \frac{1}{(1 + A\beta)} \qquad (2.13)$$

Hence the sensitivity is equal to the reciprocal of $(1 + A\beta)$. For negative feedback the magnitude of $(1 + A\beta)$ is greater than unity, so the sensitivity is less than unity. This indicates that negative feedback always gives a closed-loop gain with reduced sensitivity to variations in forward gain. This very desirable result is a general property of negative feedback. The converse result is also true for positive feedback. Using positive feedback to increase amplification requires $|1 + A\beta|$ to be less than unity and hence by Equation 2.13 the sensitivity is unfortunately increased by the same ratio.

It has already been shown that the closed-loop gain A_f becomes completely insensitive to changes in A (and is given simply by $1/\beta$) if $A\beta$ is much greater than unity. This is also shown by Equation 2.13. To make $A\beta$ much greater than unity normally requires β to be less than unity because $A \approx 1/\beta$ and is usually required to be greater than unity.

Since the feedback circuit is not required to amplify it can be wholly constructed from passive components of known values and with good stability, at low cost. The necessary high value for A is readily available because developments in integrated circuit technology have meant that cheap high-gain amplifiers (op. amps.) are manufactured in great numbers.

Instinctively it might be concluded that the reduced sensitivity to the variations in A provided by negative feedback, has to be paid for somehow. Perhaps the sensitivity to variations in the feedback fraction β is increased? It can be shown that the sensitivity to β is given by

$$S_\beta^{A_f} = \frac{\beta}{A_f} \cdot \frac{dA_f}{d\beta}$$

After some calculation, if $A\beta >> 1$,

$$S_\beta^{A_f} = \frac{-1}{(A\beta)^{-1} + 1} \approx \frac{-1}{0 + 1} = -1 \qquad (2.14)$$

This sensitivity of approximately unity means variations in β cause equal fractional variations in A_f. That is, if β has a tolerance of 1%, then provided the loop-gain is large, the tolerance in A_f also is only 1%. The benefits are paid for, of course, by using a forward amplifier with high gain and allowing this to be reduced by negative feedback to give a much lower closed loop-gain. The benefits are obtained only if $|A| >> |A_f|$.

Exercise 2.4 Prove Equation 2.14.

Differential calculus is based on small changes in variables which are made vanishingly small.

In the above sensitivity analysis the changes dA, dA_f, $d\beta$ are assumed to be infinitesimally small. In reality the parameter changes will be finite and therefore will not be exactly related by the above equations for sensitivity. Because of this the sensitivity equation is often used for the initial design of the feedback amplifier, and then an accurate assessment of the changes is made using the fundamental feedback formula. This procedure is illustrated by the following example.

Worked Example 2.2 A feedback amplifier is required to have a closed loop gain of 10 ±0.2%. The designer has available for the forward path a choice of amplifiers having gains of 50, 500 and 5000. Higher-gain amplifiers cost more than lower-gain amplifiers. Manufacturing and other tolerances cause deviations in forward gain of ±20% to occur in each type. Design the feedback arrangement to provide a stable amplifier which uses the cheapest amplifier in the forward path. Calculate the actual variations in closed-loop gain expected.

The cost is higher because more active and passive components are required to obtain higher amplification.

Solution: Applying the sensitivity Equations 2.11 and 2.13 to this, derived on the basis of infinitesimal changes in gain, cannot be an accurate procedure, but it is a simple means to estimate the required value of A.

$$0.2\% \geqslant \frac{1}{|1 + A\beta|} \times 20\%$$

that is $|1 + A\beta| \geqslant 100$.

The fundamental feedback equation (2.8), when rearranged becomes

$$A = (1 + A\beta).A_f$$

Hence $A \geqslant (100) \times (10)$ is required. Thus, $A = 5000$ is the most economical choice. Substituting this into Equation 2.8 gives

Choose the lowest amplification that does the job.

$$10 = \frac{5000}{1 + 5000\beta}$$

from which $\beta = 0.0998$. Hence the sign values are $A = 5000$ and $\beta = 0.099$.

Because this amplifier provides higher gain than the minimum required it can be expected to provide less than the $\pm 0.2\%$ variation in A_f than has been specified. According to Equations 2.11 and 2.13

$$\frac{dA_f}{A_f} = \frac{1}{1 + 5000 \times .0998} \cdot (\pm 20\%) = \pm 0.04\%$$

To calculate the actual variation in A_f note that the minimum and maximum values for A in practice are 4000 and 6000 respectively (that is, 5000 $\pm 20\%$). Substituting these extreme values into the fundamental feedback equation gives

$$A_{fmin} = \frac{4000}{1 + 4000 \times .0998} = 9.995 \text{ (.05\% low)}$$

$$A_{fmax} = \frac{6000}{1 + 6000 \times .0998} = 10.003 \text{ (.03\% high)}$$

Note that the actual deviations are not exactly those indicated by the sensitivity analysis ($\pm .04\%$). However, the design meets the specified tolerance for A_f of $\pm 0.2\%$.

Reduction of Noise and Distortion

The output of any practical amplifier as well as containing a component equal to the amplified input also contains some unwanted components. Some of these are classified as *noise* since they are present even when the input signal is removed. Sources of noise include thermal noise in active and passive components, electromagnetic interference from other circuits and residual mains-frequency hum passing along the amplifier power supply lines. Another type of unwanted output component, produced when the signal is present, is *distortion*. This is produced by non-linearities in the input-output relationship of the amplifier. That is, the gain of the amplifier varies somewhat with signal amplitude. This is owing primarily to inherent non-linearities in the characteristics of the devices in the amplifier, but also it can be caused by amplitude limiting if the signal amplitudes exceeds the range the amplifier can accommodate. In the case of distortion the actual output signal deviates from the ideal. Provided it is small, this deviation can be thought of as an

Inherently bipolar transistors have exponential characteristics and MOS field-effect transistors have square-law characteristics.

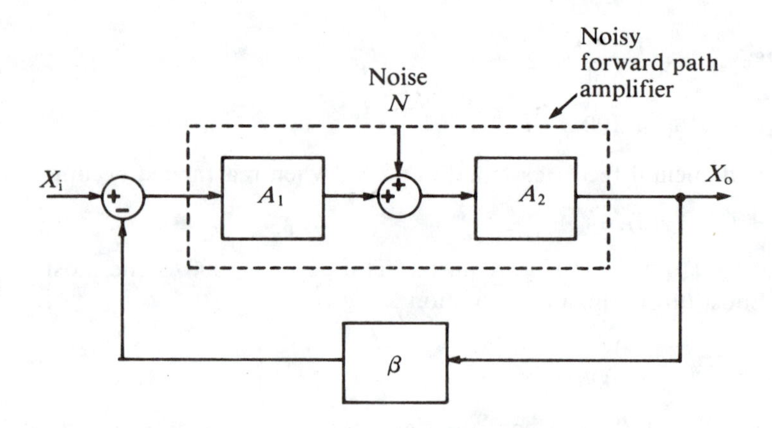

Fig. 2.5 Feedback system with noise entering forward amplifier.

unwanted component added to the signal. Distortion differs from noise in that it disappears when the signal is zero and usually increases when the signal increases, whereas noise components are generally independent of the signal amplitude.

Modelling like this is a powerful technique. It allows the essentials to be focused on.

To simplify the discussion, suppose an unwanted noise component, N, enters the amplifier somewhere between the input and output, as shown in Fig. 2.5. The block A_1 represents the gain of A before the entry of N and block A_2 represents the gain after the entry of N. Therefore $A = A_1A_2$. If N occurs towards the input end of the amplifier, A_1 is low, and if N occurs towards the output end of the amplifier, A_2 is low. To analyse the feedback configuration, first note that

$$\begin{aligned} X_o &= A_2 \times \{\text{output of block } A_1 + N\} \\ &= A_2\{A_1(X_i - \beta X_o) + N\} \end{aligned}$$

That is,

$$(1 + A_1A_2\beta)X_o = A_1A_2X_i + A_2N$$

Finally, making use of $A = A_1A_2$,

$$X_o = \frac{A_1A_2}{1 + A\beta} \cdot X_i + \frac{A_2}{1 + A\beta} N \qquad (2.15)$$

The second term on the right-hand side of this expression gives the unwanted component at the output after feedback is applied.

Signal-to-noise ratio is widely used in other branches of electronics and also communications theory. High signal-to-noise ratio is desired.

Consider the *signal-to-noise ratio* (S/N) at the output. This is defined as the ratio of the wanted to unwanted signal components. Using Equation 2.15

$$\frac{S}{N} = \frac{A_1A_2}{1 + A\beta} \cdot X_i \Bigg/ \frac{A_2}{1 + A\beta} \cdot N$$

That is

$$\frac{S}{N} = A_1 \cdot \frac{X_i}{N} \qquad (2.16)$$

This relationship shows that the effect of noise N entering the system becomes less serious if A_1 is larger (that is, more gain present between the input point and the noise source). The worst case occurs if the noise enters at the comparator so that it is

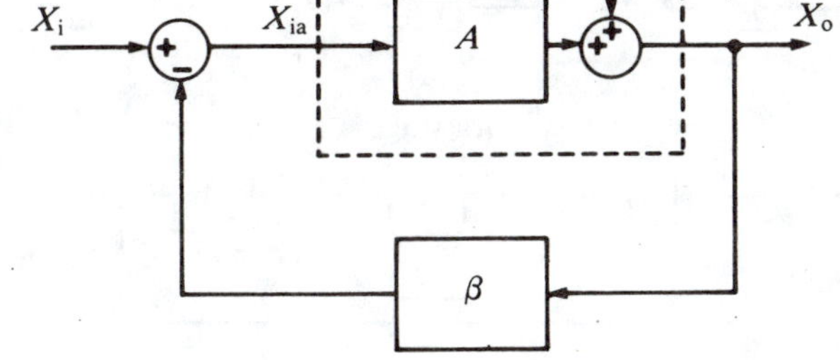

Fig. 2.6 Feedback system with output distortion.

added directly onto the signal since no amount of feedback can then reduce the noise relative to the signal. In this case $A_1 = 1$ and the same signal-to-noise ratio exists everywhere in the system. For this reason, in the design of high-performance amplifiers, care must be taken to use low-noise components in the early stages of the amplifier and to shield such stages from interference.

Distortion.

The output stage of an amplifier is frequently designed to give as much output power as possible, the main limitation being the distortion which arises if the output transistors are driven too hard. Negative feedback enables distortion to be reduced for the same output power. Fig. 2.6 shows the basic feedback amplifier with distortion represented by D added. It should be noted that the distortion signal is a function of signal level X_o, but is fixed for a given X_o.

For this arrangement

$$X_{ia} = X_i - \beta X_o = X_i - \beta (A X_{ia} + D)$$

Eliminating X_{ia} gives

$$X_o = \frac{A}{1 + A\beta} \cdot X_i + \frac{D}{1 + A\beta} \qquad (2.17)$$

Check this.

For a given output signal X_o, distortion D is fixed but its effective magnitude at the output has been reduced by $1 + A\beta$. This is a significant advantage. At the same time either the input signal has to be increased or the gain A has to be increased (e.g. by adding a pre-amplifier which operates at low signal power and need not introduce significant distortion). Thus, power amplifiers can be made with very low distortion despite the non-linear characteristics of the transistors.

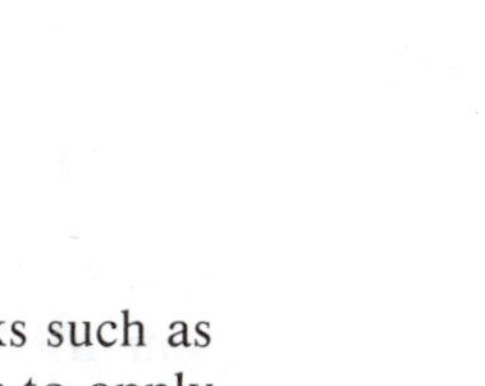

Multiple-Loop Feedback

Often the forward amplifier block is made from a chain of smaller blocks such as individual transistor stages. The question then arises; whether it is better to apply individual feedback loops around each block to make a *multiple-loop system*, or to apply an overall single feedback loop?

Suppose for simplicity that there are N identical blocks; then the two situations

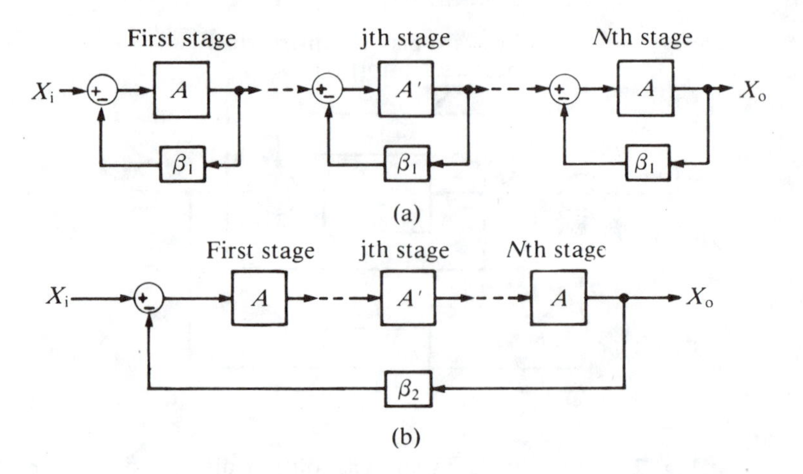

Fig. 2.7 Feedback applied to several stages of amplification: (a) feedback applied to each stage, (b) feedback applied overall.

are illustrated in Fig. 2.7. To make a fair comparison it is assumed that the closed-loop gains in both cases are nominally the same. Consider the comparison of the two cases from the viewpoint of sensitivity to changes in one of the amplifier gains. To maintain generality suppose this amplifier is at the jth position and has gain A' which is nominally equal to A but changes for some reason such as component variations.

For the first case, Fig. 2.7a, the overall closed-loop gain A_{fa} is given by the product of the closed-loop gain for each block,

$$A_{fa} = \frac{A}{1 + A\beta_1} \times \dots \times \frac{A'}{1 + A'\beta_1} \times \dots \times \frac{A}{1 + A\beta_1}$$

$$= \frac{A^{N-1}}{(1 + A\beta_1)^{N-1}} \cdot \frac{A'}{1 + A'\beta_1} \tag{2.18}$$

Applying the sensitivity definition gives, using the quotient rule for differentiation,

$$S_{A'}^{A_{fa}} = \frac{A'}{A_{fa}} \cdot \frac{dA_{fa}}{dA'} = \frac{A'}{A_{fa}} \cdot \frac{A^{N-1}}{(1 + A\beta_1)^{N-1}} \cdot \frac{(1 + A'\beta)(1) - A'(\beta_1)}{(1 + A'\beta_1)}$$

$$= \frac{A'}{A_{fa}} \cdot \frac{A^{N-1}}{(1 + A\beta_1)^{N-1}} \frac{1}{(1 + A'\beta_1)^2}$$

Using Equation 2.18 on this expression gives

Verify this yourself.

$$S_{A'}^{A_{fa}} = \frac{A_{fa}}{A'} \tag{2.19}$$

Now looking at the second case, Fig. 2.7b, of feedback applied over all the amplifiers, the effective forward gain is $A^{N-1}A'$ and so the overall closed-loop gain is,

β for the second case is different from the first case if equal closed-loop gains are to be obtained.

$$A_{fb} = \frac{A^{N-1}A'}{1 + A^{N-1}A'\beta_2} \tag{2.20}$$

Again applying the sensitivity definitions, we have

$$S_{A'} = \frac{A'}{A_{fb}} \cdot \frac{dA_{fa}}{dA'} = \frac{A'}{A_{fb}} \cdot \frac{A^{N-1}}{(1 + A^{N-1}A'\beta_2)^2}$$

Confirm the differentiation yourself.

Using Equation 2.20 this time gives,

$$S_{A'}^{A_{fb}} = \frac{A_{fb}}{A^{N-1}A'} \tag{2.21}$$

Combining the two sensitivity equations (2.19 and 2.21) and also noting that for equal closed-loop gain for both systems, $A_{fa} = A_{fb}$, gives

$$S_{A'}^{A_{fb}} = \frac{1}{A^{N-1}} \cdot S_{A'}^{A_{fa}} \tag{2.22}$$

This clearly shows that the sensitivity of the single loop arrangement in Fig. 2.7b is better than the first because $A > 1$, and A^{N-1} is even larger than this.

It can be shown that the single loop arrangement is better in other respects, such as more nearly ideal closed-loop input and output impedances (the effect of feedback on input and output impedance is examined later in Chapter 4). The main problem in following this approach is that the individual amplifier phase-shifts add up and are more likely to give positive feedback which can lead to instability. This places a limit on the number of stages around which a single feedback loop is applied, and typically a maximum of two or three stages are used for the forward amplifier.

Good news and bad news. Engineering often involves a balance of advantages and disadvantages.

The analysis of the multiple-loop system in Fig. 2.7a presents no real problem because the system is readily recognized to consist of a cascade of single-loops. In other multiple-loop systems it is not so easy. An example is the system shown in Fig. 2.9a. To simplify the analysis of these more complicated systems the use of transformations can be helpful. A small collection of these is shown in Fig. 2.8.

Exercise 2.5

Prove the transformations shown in Fig. 2.8.

This is easily done by writing down the input-output equations for both sides of the transformation and showing they are equivalent.

The use of these transformations is indicated by the following example

---

Worked Example 2.3

Derive the overall gain between X_i and X_o in Fig. 2.9a.

Solution. One way is to recognize first that transformation (c) in Fig. 2.8 can be used on blocks C and B to give the result shown in Fig. 2.9b, so that the input to C is moved to the other side of B, thus placing A and B directly in cascade.

On this figure it can be seen that blocks A, B and E now constitute a normal single loop feedback arrangement and can therefore be replaced by a single block whose gain is given by the fundamental feedback equation, as shown in Fig. 2.9c. Also blocks F and C/B can be combined using the transformation Fig. 2.8b noting the differing signs at the comparators. The result of this operation is shown in Fig. 2.9c.

This figure is recognized as a single loop and again using the fundamental feedback equation the final result is

$$\text{Overall gain} = \frac{D.\dfrac{(AB)}{1 + ABE}}{1 + D.\dfrac{(AB)}{1 + ABE} \cdot (F - \dfrac{C}{B})}$$

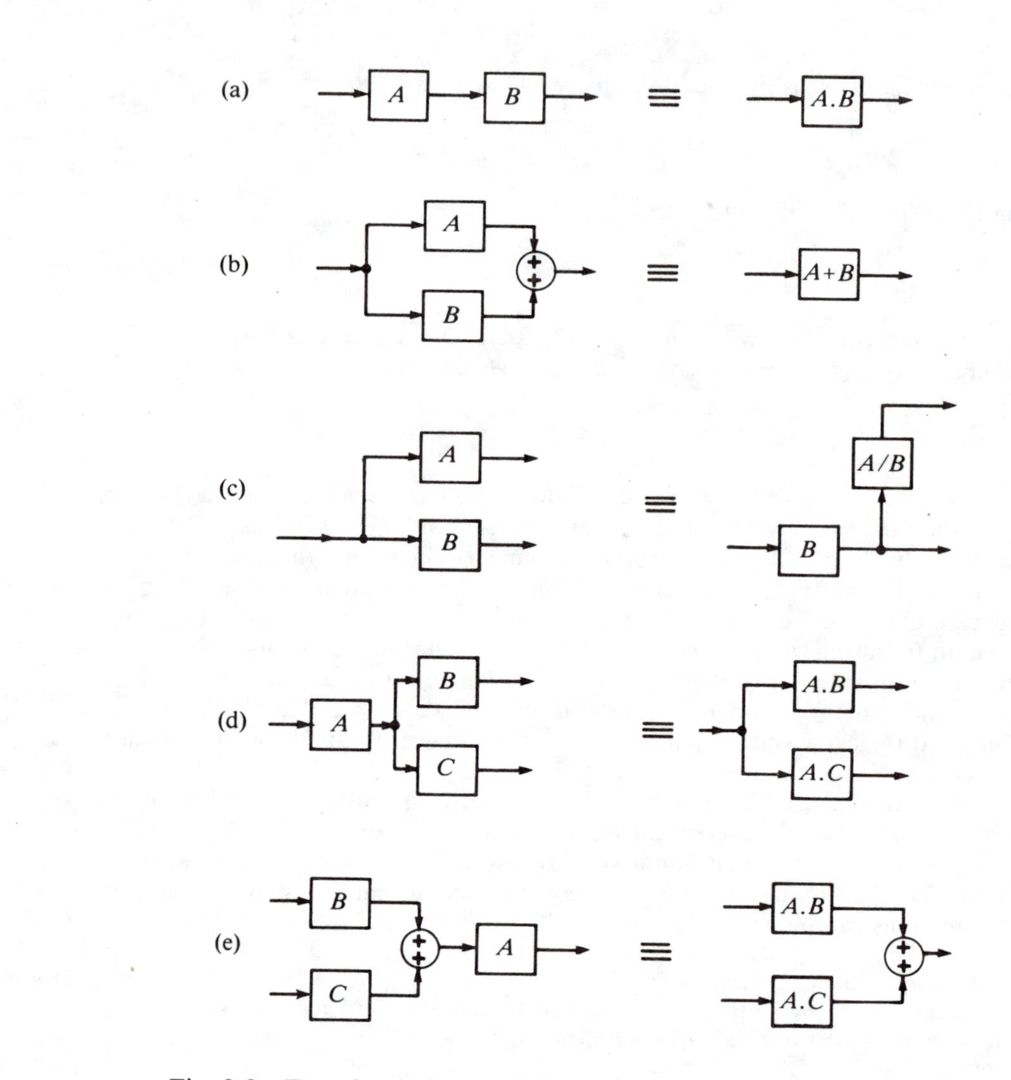

Fig. 2.8 Transformations to help analyse multiple-loop feedback systems.

Moschytz, G.G. *Linear Integrated Networks: Fundamentals* (Van Nostrand Reinhold, 1974) has a good account of signal-flow graphs.

In many cases the use of simple transformations such as those given in Fig. 2.8 allows the performance to be calculated. Sometimes examples may occur where more formal methods have to be used (such as those based on signal-flow graphs) but these methods are outside the scope of this text.

Summary

The feedback system comprises a *forward-path amplifier* A, a *feedback block* β, and a *comparator*. The resulting *closed-loop gain* A_f obtained as a result of feedback can be calculated from the *fundamental feedback formula* $A/(1 + A\beta)$. The type of feedback obtained, *negative* or *positive*, depends on the value of $|1 + A\beta|$. If the feedback is positive then the forward-path amplifier gain is increased so that A_f is greater than A; if the feedback is negative then the resulting A_f is less than A.

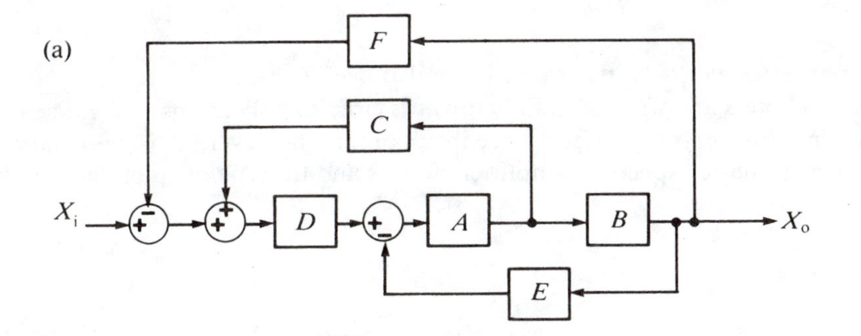

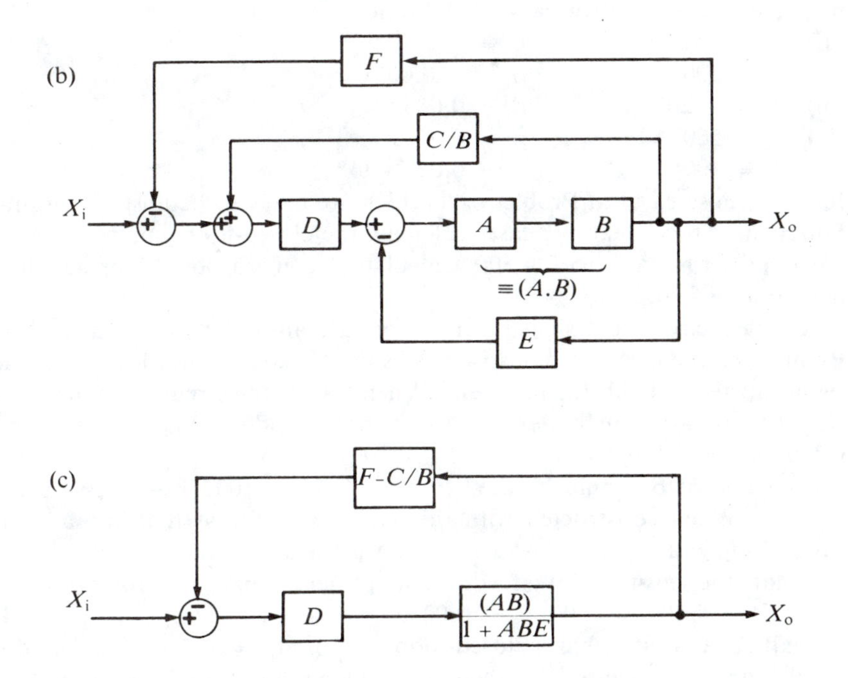

Fig. 2.9 Simplification of multiple-loop feedback system.

Despite the reduction in gain caused by negative feedback this type of feedback is usually preferred in the design of amplifiers because important benefits are obtained. One of these is the *reduced sensitivity* of the closed-loop gain to variations in forward-path amplification which depends on the characteristic of active devices, particularly transistors, which are not constant. If the forward-path gain is made very high the system is practically insensitive to gain variations and the closed-loop gain in the limit is defined entirely by the feedback block (becoming $\approx 1/\beta$). This is of great practical significance because the feedback circuit can be made cheaply and accurately from passive components. The necessary high forward-path amplification can be obtained cheaply because of the advances of modern integrated circuit technology.

We have also seen that the effect of *noise* is worse in any amplifier where it occurs early in the amplifier. Negative feedback also reduces the *distortion* in the output of an amplifier operating at a given level of output signal amplitude.

Properties of *multi-loop systems* have also been considered.

Later chapters are oriented mainly towards circuit applications of feedback and to developing further properties of feedback. First, however, it is necessary to be familiar with basic aspects of amplifier circuits and the next chapter deals with this.

Problems

2.1 For the system shown in Fig. 2.1, determine the type of feedback (that is, negative or positive) for each of the following cases:

(i)	$A = -100$	$\beta = 1\%$
(ii)	$A = 200$	$\beta = -0.0051$
(iii)	$A = -200$	$\beta = 0.0049$
(iv)	$A = 500$	$\beta = 0.01$
(v)	$A = 400\underline{/10^\circ}$	$\beta = 0.02\underline{/-15^\circ}$

2.2 Identify those cases in Problem 2.1 which are likely to be unstable in practice. For each of the remaining cases calculate the closed-loop gain.

2.3 An amplifier has a gain of -500. Calculate β to give a closed-loop gain having a magnitude of 10.

2.4 A feedback circuit is designed using a forward amplifier with gain $A = +100$. By mistake the circuit is constructed using a comparator which adds the two signals instead of subtracting them. When tested, the circuit is found to have a closed-loop gain which is exactly twice that intended. What was the intended value of closed-loop gain?

2.5 What value of β should be used to give $A_f = -100$, if A is assumed to be infinity? When constructed with this value of β the system had a measured closed-loop gain of $A_f = -99.9$; what was the actual value of A?

2.6 Calculate the sensitivity of a feedback amplifier to changes in forward-amplifier gain, given that $A = 800$, $\beta = 0.025$.

2.7 A feedback amplifier has a closed-loop gain of $A_f = +200$. It is found that a small change in A causes a change in A_f which is a twentieth of that in A. What is the value of A?

2.8 A negative feedback amplifier has $A = 1000$, $\beta = 0.02$. With an input of $X_i = 10^{-3}$ V r.m.s., the measured signal-to-noise ratio at the output is 100. Assuming the noise is due to hum what is the output-noise r.m.s. amplitude if the input signal is removed?

Amplifiers without Feedback 3

Objectives

- ☐ To describe the basic properties of amplifiers without feedback.
- ☐ To describe the four types of amplifier model, the equivalence that exists between them, and which model to choose in practice.
- ☐ To quantify the effect of input and output impedance on amplifier performance, and to couple signal source, amplifiers and load to achieve good coupling of signal voltage, current, or power.
- ☐ To explain the main causes of the reduction in amplifier gain at high and low frequencies and to be able to calculate their effects.
- ☐ To describe the function of a differential amplifier and to be able to distinguish between common-mode and differential mode quantities. Explain why a differential amplifier can be used to amplify signals which are subject to interference.

Amplifier Models

The general properties of feedback and some of its advantages were considered in Chapter 2. This chapter looks at the fundamentals of amplifiers. It is necessary to do this before considering how to apply feedback to electronic circuits. The emphasis is on the terminal behaviour of amplifiers rather than looking in detail at the various ways in which amplifiers are constructed from active devices such as BJTs and FETs.

BJT and FET amplifying circuits are covered in Ritchie, G.J. *Transistor Circuit Techniques*, Second edition, Van Nostrand Reinhold (International), 1987.

The basic connection of signal source, amplifier and load is shown in Fig. 3.1. The signal power in the load is normally greater than the power provided by the signal source. Also power in the form of heat is dissipated in the amplifier. These extra powers come from a power supply as indicated in the figure which normally provides energy at one or more constant voltage levels, either from batteries or from rectified and smoothed mains supply. In effect the amplifier converts this power from constant voltages into a varying output voltage v_{out} which follows the input signal, v_{in}. In subsequent figures the supply is omitted for simplicity.

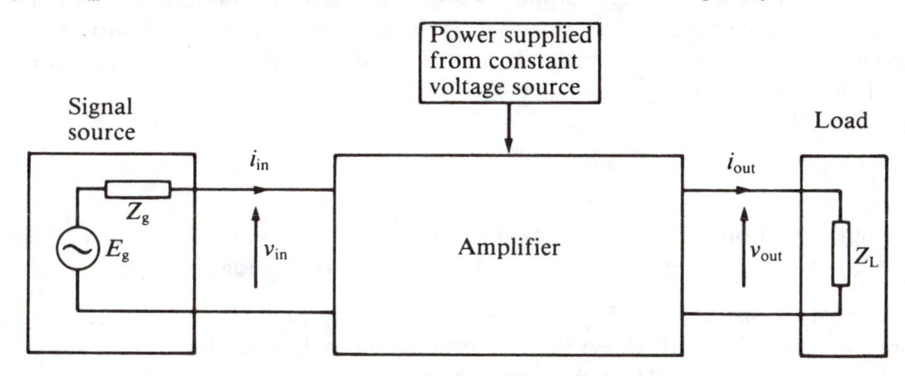

Fig. 3.1 Amplifier with signal source and load.

The amplification, or gain, of the amplifier may be looked at from the point of view of either voltage or current signals. Voltage and current gain are defined as follows:

$$\text{Voltage gain, } A_V = \frac{\text{Output signal voltage}}{\text{Input signal voltage}} = \frac{v_{out}}{v_{in}} \tag{3.1}$$

$$\text{Current gain, } A_I = \frac{\text{Output signal current}}{\text{Input signal current}} = \frac{i_{out}}{i_{in}} \tag{3.2}$$

The quantities A_V, A_I are dimensionless since they are the ratios of signals measured in the same units. Two more definitions of gain are obtained by choosing signals measured in different units. These are

$$\text{Transfer Resistance, } R_T = \frac{\text{Output signal voltage}}{\text{Input signal current}} = \frac{v_{out}}{i_{in}} \text{ ohms} \tag{3.3}$$

$$\text{Transfer Conductance, } G_T = \frac{\text{Output signal current}}{\text{Input signal voltage}} = \frac{i_{out}}{v_{in}} \text{ siemens} \tag{3.4}$$

Although R_T and G_T have units of ohms and siemens they do not necessarily correspond to a single defining resistive component in the circuit. In general they will be functions of several component parameters within the amplifier.

Suppose we wish to choose an amplifier to increase the current-measuring sensitivity of an ammeter. The first requirement is that the current gain A_I of the amplifier should equal the factor by which the ammeter sensitivity is to be increased. (Typically this would be a power of ten since it would allow the existing scale graduations on the ammeter to be used.) In practice, the amplifier-ammeter combination is used in the same way as an ordinary ammeter. That is, the path containing the current to be measured is broken and then joined to the amplifier input terminals, so the unknown current flows through the amplifier input. As with an ordinary ammeter, the amplifier should disturb as little as possible the circuit containing the current to be measured (the amplifier input voltage should be as small as possible, and ideally zero). So a perfect amplifier for this application would have an input which behaves as a short-circuit. At the output of the amplifier the output current passes through the ammeter. The ammeter which constitutes the amplifier load has a resistance which is small but not zero. To maintain accuracy, it is desirable that this output current is not affected by whatever voltage is developed across the load. Ideally, therefore, the amplifier output should behave as a perfect current generator. These conditions define the *ideal current amplifier*, as depicted in Fig. 3.2a. The generator current has a value which depends on the input current and because of this the ideal current amplifier is also called a *current-controlled current-source* (*c.c.c.s.*). The c.c.c.s. parameter K_I is the constant of proportionality relating the controlling current to the controlled current. Obviously, in this case the current gain A_I is equal to K_I. In summary, the following conditions apply to an ideal current amplifier:

The load here is an ammeter so the output voltage developed across the load by i_{out} would be quite small, typically a few tenths of a volt. Other types of load could well result in higher voltages.

Input condition	$v_{in} = 0$
Transfer condition	$i_{out} = K_I\, i_{in},\ A_I = K_I$
Output condition	i_{out} is constant with respect to variations in v_{out}, i.e. independent of the resistance of the load

These conditions are expressed in graphical form in Fig. 3.3a. A comparison of these characteristics with those of a common-emitter bipolar junction transistor, Fig. 3.3b shows that the BJT behaves somewhat like an ideal current amplifier. The approximation is not perfect, as the BJT input voltage V_{BE} is not zero (Fig. 3.3b),

The FET characteristics also are a set of nearly horizontal lines, but being controlled by a voltage (the gate-source voltage) the FET more nearly approximates to an ideal transconductance amplifier (see below).

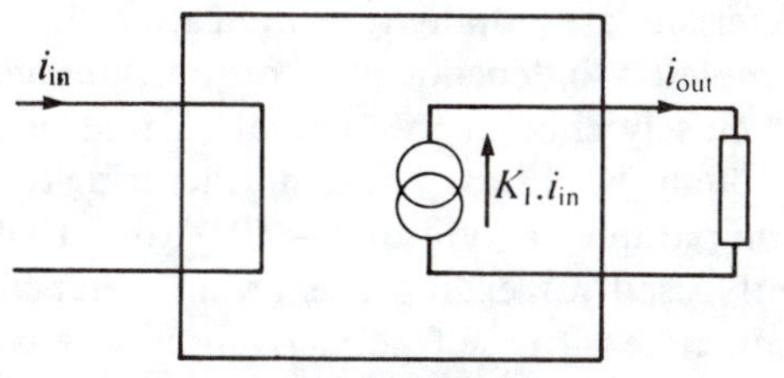

(a) Ideal current amplifier
(current-controlled current source)

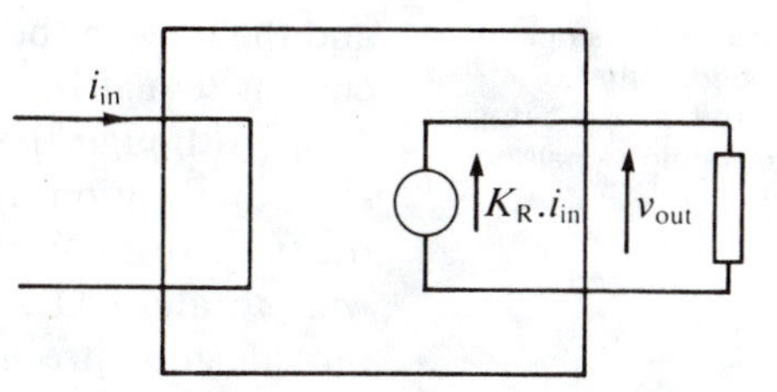

(b) Ideal trans-resistance amplifier
(current-controlled voltage source)

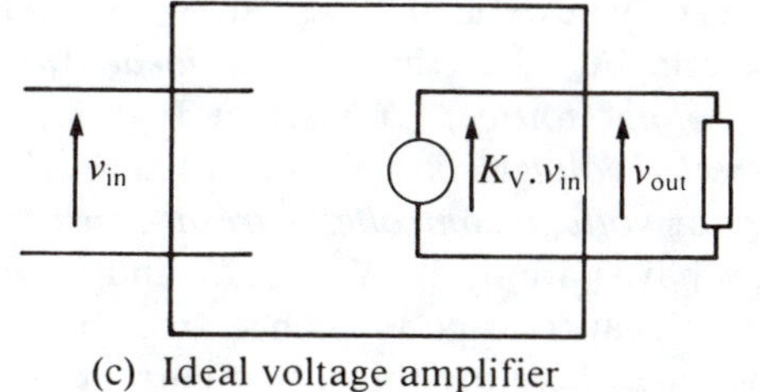

(c) Ideal voltage amplifier
(voltage-controlled voltage source)

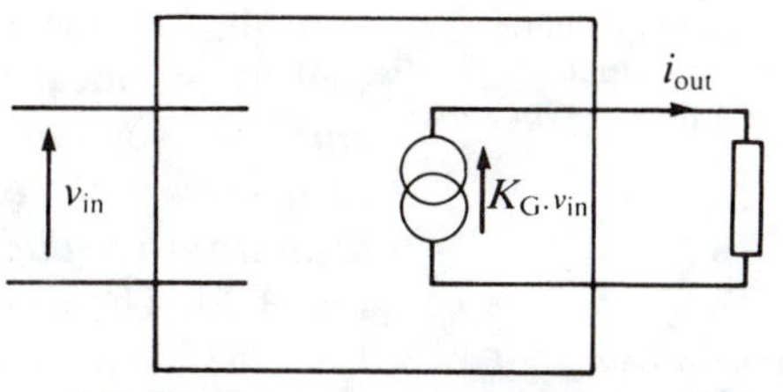

(d) Ideal trans-conductance amplifier
(voltage-controlled current source)

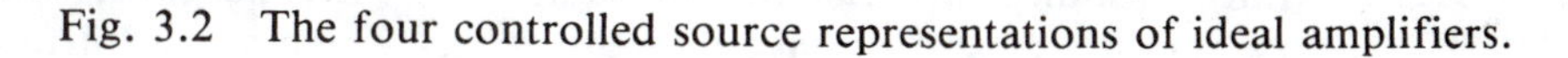

Fig. 3.2 The four controlled source representations of ideal amplifiers.

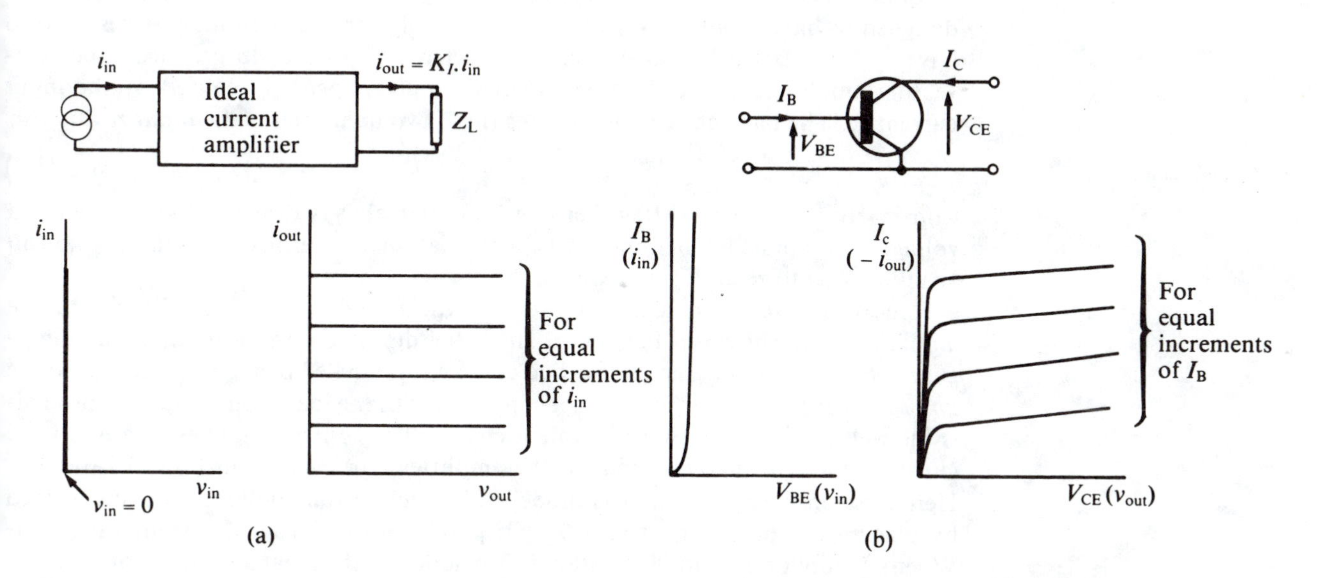

Fig. 3.3 Current amplifier characteristics: (a) ideal, (b) bipolar junction transistor.

A practical transistor amplifier circuit also needs other components to bias the circuit to allow alternating signals to be amplified.

and the lines in the output characteristic are sloping thus indicating that the output current (equal to $-I_C$) is not completely independent of output voltage (equal to V_{CE}). Although the BJT behaves closely enough to an ideal current generator for some applications it is not good enough for others. The manufacturing tolerances on the transistor current amplification parameter (typically -50% to $+100\%$) are too wide to allow it to be conveniently used for example as an ammeter amplifier to provide good precision. Modern practice is to use feedback amplifier techniques for this application and it is shown later that the use of feedback overcomes the problem of tolerances in the current gain of transistors.

Exercise 3.1 Show that for the ideal current amplifier, $A_I = K_I$, $A_V = \infty$, $G_T = \infty$, and $R_T = K_I.R_L$ where R_L is the load resistance.

Two types of amplifier input and two types of output, therefore four amplifier types in all.

In all, four ideal amplifier models may be defined. These are shown in Fig. 3.2. As well as the ideal current amplifier in Fig. 3.2, there is the *ideal trans-resistance amplifier* (or *current-controlled voltage-source*, c.c.v.s.) in Fig. 3.2b; the *ideal voltage amplifier* (or *voltage-controlled voltage-source*, v.c.v.s.) in Fig. 3.2c; and the *ideal trans-conductance amplifier* (or *voltage-controlled current-source*, v.c.c.s.) in Fig. 3.2d. These controlled sources have parameters K_I, K_R, K_V and K_G respectively. Like the ideal current amplifier the ideal trans-conductance amplifier has an ideal current generator at the output side and therefore the output current is not affected by the output voltage developed across the load. The other two ideal amplifiers (the trans-resistance amplifier and the voltage amplifier) have ideal voltage generators at the output side and so have output voltages which are not affected by the output currents flowing through the loads. At the input sides, two of the ideal amplifiers (the current amplifier and trans-resistance amplifier) develop no voltage at the input terminals and therefore have perfect current sensing; while the other two ideal amplifiers have zero input currents and therefore have perfect voltage sensing.

For the ideal amplifiers, A_I, R_T, A_V, and G_T are numerically equal to K_I, K_R, K_V, and K_G respectively. However, this is not necessarily so when loading effects in non-ideal amplifiers are considered later.

The four types of controlled sources serve as useful ideals, but in modelling real amplifiers account must be taken of the fact that amplifier circuits cannot be designed to have inputs in which v_{in}, or i_{in} is truly zero nor can they be designed to have v_{out}, or i_{out} which is absolutely unaffected by the load. In practice, a current-sensing amplifier always develops a small input voltage, v_{in}, caused by the input current i_{in}. The parameter which relates these two quantities is the *input resistance*,

Ohm's law once again.

$$r_{in} = v_{in}/i_{in} \tag{3.5}$$

Conversely a practical voltage-sensing amplifier always draws a finite i_{in} when a voltage v_{in} is applied. The input resistance relationship defined in Equation 3.5 can again be used to relate v_{in} and i_{in}.

Non-ideal behaviour is also present at the output of real amplifiers. In the amplifier types shown in Figs. 3.2b and c the dependence in practice of a voltage generator terminal voltage on the load current is modelled by a resistance, called the *amplifier output resistance* r_{out} placed in series with the ideal voltage generator. This is the well known Thevenin equivalent circuit. The converse applies to a non-ideal current generator at the output in the amplifier types shown in Figs. 3.2a and d. Here the non-ideal behaviour is modelled by the Norton equivalent circuit formed by placing the output resistance r_{out} in parallel with the ideal-current generator. When the input and output resistances are added to the ideal amplifiers of Fig. 3.2, then the circuits which model the real behaviour of amplifiers are those shown in Fig. 3.4.

When sinusoidal signals are applied to amplifiers then phase angles are usually observed between the various voltages and currents and so the input and output resistances must be replaced by complex input and output impedances, Z_{in} and Z_{out}.

Exercise 3.2

An ideal voltage amplifier has $Z_{in} = \infty$ ohms, and $Z_{out} = 0$ ohms. What are the input and output impedances of the other three ideal amplifiers?

The input sides of the four amplifier representations in Fig. 3.4 are the same. Each contains just an input resistance r_{in}. Although the circuits differ in that two of them have generators controlled by the voltage across r_{in} and the other two are controlled by the current through r_{in}. This has no effect on the amplitudes of v_{in} and i_{in} which are determined by the generator applied to the amplifier and are in a fixed ratio. The output sides of the four amplifiers contain either Thevenin or Norton equivalent circuits, and we know from basic circuit theory that a Norton equivalent circuit can be transformed to a Thevenin equivalent circuit, or vice versa. It might be wondered therefore if it is possible to transform from one amplifier model to another? Consider for example the terminal equations for the c.c.c.s.-based amplifier model (Fig. 3.4a) and the v.c.v.s.-based amplifier model (Fig. 3.4c).

A Thevenin equivalent of voltage generator E in series with resistance R has the identical terminal behaviour to a Norton equivalent of a current generator J in parallel with a resistance, provided this resistance also equals R and $E = J.R$.

For the c.c.c.s.-based model:

Amplifier input $\quad v_{in} = i_{in}.r_{in} \quad$ (3.6)

Amplifier output $\quad i_{out} = K_I.i_{in} - \dfrac{v_{out}}{r_{out}} \quad$ (3.7)

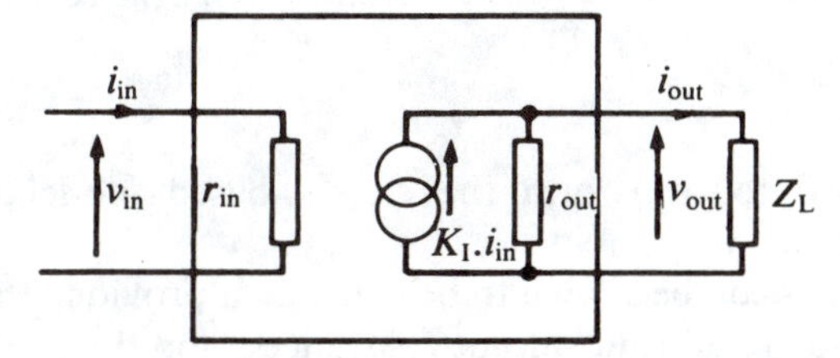

(a) Real current amplifier

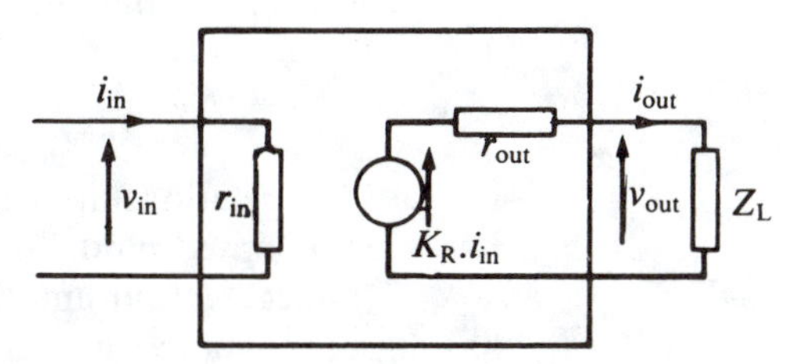

(b) Real trans-resistance amplifier

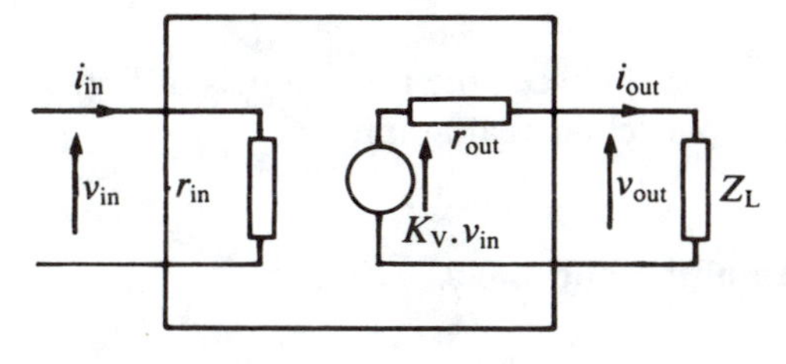

(c) Real voltage amplifier

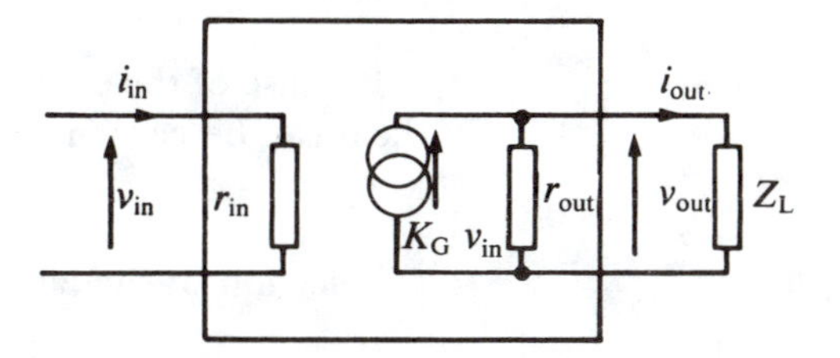

(d) Real trans-conductance amplifier

Fig. 3.4 The four ways to model real amplifiers.

The equation for the amplifier input (3.6) is the Ohm's law relationship for the input resistance. Equation 3.7 for the output voltage is obtained by observing that in Fig. 3.4a the amplifier output current is equal to the current from the current generator $K_I \, i_{in}$ minus the current which flows through the resistance r_{out}, which by Ohm's law equals v_{out} divided by r_{out}.

For the v.c.v.s.-based model:

$$\text{Amplifier input} \quad v_{in} = i_{in}.r_{in} \tag{3.8}$$

$$\text{Amplifier output} \quad v_{out} = K_V.v_{in} - i_{out}.r_{out} \tag{3.9}$$

Here the output voltage is given by the generator voltage $K_V.v_{in}$ minus the voltage drop $i_{out}.r_{out}$ across the output resistance.

It can be seen that the input equations (3.6 and 3.8) for both circuits are identical. Turning now to the output equation (3.9) for the v.c.v.s.-based model, after rearranging this becomes

The strategy is to make the equation have the same form as Equation 3.7; i_{out} on the left and i_{in} and v_{out} on the right.

$$i_{out} = \frac{K_V}{r_{out}}. v_{in} - \frac{v_{out}}{r_{out}}$$

and substituting for v_{in}, using Equation 3.8 gives

$$\text{Amplifier output} \quad i_{out} = K_V \frac{r_{in}}{v_{out}} i_{in} - \frac{v_{out}}{r_{out}} \tag{3.10}$$

A comparison of the input and output equations for the v.c.v.s.-based model (3.8 and 3.10) with the corresponding equations for the c.c.c.s.-based model (3.6 and 3.7) reveals that both models are fully equivalent provided both input and output resistances are the same and the control source gain constants satisfy the relationship

$$K_I = K_V . \frac{r_{in}}{r_{out}} \tag{3.11}$$

This relationship can be used to transform the c.c.c.s.-based model to the v.c.v.s.-based model or vice versa.

In fact the four amplifier models can be shown to be equivalent provided the four input resistances are equal and so are all four output resistances, and the controlled source parameters satisfy the particular relationship Equation 3.12 below.

Exercise 3.3 Show that the four amplifier models in Fig. 3.4 are equivalent provided the controlled source parameters satisfy the relationship

$$K_I = K_V . \frac{r_{in}}{r_{out}} = K_R . \frac{1}{r_{out}} = K_G.r_{in} \tag{3.12}$$

Because of the equivalence that exists between the four models of Fig. 3.4, any of them can be chosen to represent a particular real amplifier.

Connection of Signal-Source, Amplifier and Load

The calculation of the output signal of an ideal amplifier (Fig. 3.2) is a simple operation; the input signal is multiplied by the amplifier gain. If several amplifiers are connected in a chain to make a single amplifier then the overall gain is the

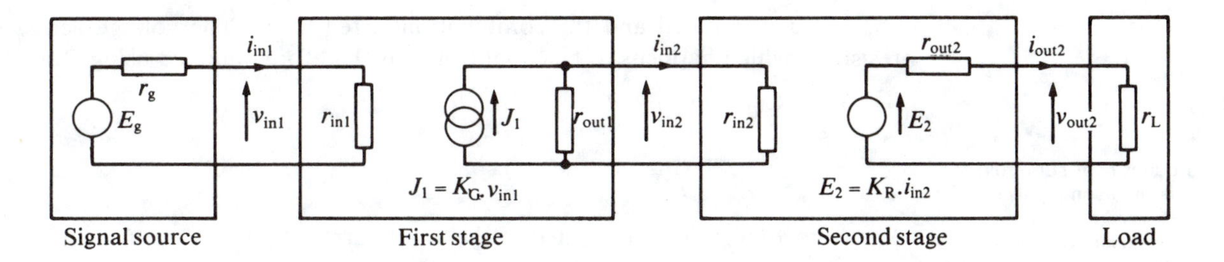

Fig. 3.5 Two-stage amplifier example.

product of the gains of the individual amplifiers. However, in practice the presence of the finite input and output resistances of the amplifiers must be taken into account when calculating output signals.

In a *chain* (or *cascade*) of amplifiers one amplifier output is connected to the input of the next, and so on. The first input and last output are then those of the overall amplifier. Amplifiers are cascaded to obtain higher amplification.

The way this is done can be seen by considering the two-stage amplifier arrangement shown in Fig. 3.5. The first amplifier stage is a trans-conductance amplifier and the second a trans-resistance amplifier. Although this is just one of the many possible arrangements that can be made using the four basic types of amplifier (see Fig. 3.4), this example illustrates points which are generally applicable.

The signal is applied from some generator having e.m.f. E_g and self-resistance r_g. This is connected to the first amplifier which is of the voltage-controlled type. Hence it is the signal voltage rather than current that is the significant factor at this point in the circuit. To determine the actual voltage applied to the amplifier first note that the generator E_g and the two resistances r_g and r_{in_1} form a single loop. The current flowing in this loop, which equals i_{in_1}, is given by Ohm's law:

$$i_{in_1} = \frac{E_g}{r_g + r_{in_1}} \tag{3.13}$$

Also $v_{in_1} = r_{in_1}.i_{in_1}$ and therefore

$$v_{in_1} = E_g . \frac{r_{in_1}}{r_g + r_{in_1}} \tag{3.14}$$

The term to the right of E_g in this expression is known as the *voltage coupling factor*. The coupling factor is less than unity, since the effect of r_g and r_{in_1} is to cause the voltage at the generator terminals to be reduced when connection is made to the amplifier. However, it is possible to have a coupling factor which is close to the maximum value of unity by ensuring that r_{in_1} is much greater than r_g.

Often the amplifier designer has no control over the value of r_g, which is a characteristic of the transducer generating the electrical signal input. To obtain a good coupling factor, an amplifier is chosen having an input resistance which is much greater than r_g. In practice a value of r_{in} ten or more times that of r_g is usually satisfactory.

It usually happens that an engineer has to accept some constraints on a problem. The task is to solve the problem within the constraints.

The voltage v_{in_1} controls the current generator at the ouput of the first stage, $J_1 = K_G.v_{in_1}$. This output is connected to the input of the second stage, which is current controlled by i_{in_2}, and the signal currents, rather than signal voltages, are being coupled at this point. It can be seen that some of the current J_1 flows into r_{out_1} and hence not all of it is coupled to the second amplifier input. If the boxes drawn round the amplifiers are disregarded for the moment it is seen that r_{out_1} and r_{in_2} are connected in parallel and the combination is fed by J_1. The voltage across these resistors, which happens to be equal to v_{in_2}, is therefore obtained using Ohm's law:

The 'product over sum' rule for two resistances in parallel.

$$v_{in_2} = (r_{out_1} \| r_{in_2}).J_1$$

$$= \frac{r_{out_1}.r_{in_2}}{r_{out_1} + r_{in_2}} . J_1 \tag{3.15}$$

Readers familiar with electrical circuit theory will recognize the current divider principle here and the voltage divider principle in Equation 3.14.

The voltage v_{in_2} in turn is related to the input current by $v_{in_2} = r_{in_2}.i_{in_2}$, and so

$$i_{in_2} = \frac{v_{in_2}}{r_{in_2}} = \frac{1}{r_{in_2}} . \frac{r_{out_1}.r_{in_2}}{r_{out_1} + r_{in_2}} J_1 = J_1 \frac{r_{out_1}}{r_{out_1} + r_{in_2}} \tag{3.16}$$

The term on the right of J_1 in this expression is the *current coupling factor*. Like the voltage coupling factor, it is generally less than unity. It can be made close to unity by choosing r_{in_2} to be much smaller than r_{out_1}.

The input current to the second stage controls the output voltage generator, $E_2 = K_R.i_{in_1}$, which provides the load voltage, v_{out_2}. Here the signal voltages are coupled. A similar calculation to that done for the voltage coupling at the input reveals

$$\text{Output voltage coupling factor} = \frac{v_{out_2}}{E_2} = \frac{r_L}{r_{out_2} + r_L} \tag{3.17}$$

It is readily seen that in this analysis the quantities r_g, r_{out_1} and r_{out_2} are the effective self-resistances of the three generators in Fig. 3.5, while the quantities r_{in_1}, r_{in_2} and r_L are respectively the effective load resistances connected to the three generators. Therefore Equations 3.15 to 3.17 and the accompanying discussion can be stated in the general form

$$\text{Voltage coupling factor} = \frac{\text{Effective load resistance}}{\text{Effective source resistance} + \text{Effective load resistance}} \tag{3.18}$$

$$\approx 1$$

provided (Effective source resistance) $<<$ (Effective load resistance).

Be careful to distinguish the differences in these rather similar looking equations and conditions.

$$\text{Current coupling factor} = \frac{\text{Effective source resistance}}{\text{Effective source resistance} + \text{Effective load resistance}} \tag{3.19}$$

$$\approx 1$$

provided (Effective source resistance) $>>$ (Effective load resistance).

Putting the results of the above analysis together, the output voltage for Fig. 3.5 is given by the expression

$$v_{out_2} = E_g \cdot \frac{r_{in_1}}{r_g + r_{in_1}} \cdot K_G \cdot \frac{r_{out_1}}{r_{out_1} + r_{in_2}} \cdot K_R \cdot \frac{r_L}{r_{out_2} + r_L} \tag{3.20}$$

$$= E_g \cdot \begin{pmatrix}\text{Input voltage} \\ \text{coupling factor}\end{pmatrix} \cdot K_G \cdot \begin{pmatrix}\text{Inter-stage current} \\ \text{coupling factor}\end{pmatrix} \cdot K_R \cdot \begin{pmatrix}\text{Output voltage} \\ \text{coupling factor}\end{pmatrix} \tag{3.21}$$

Worked Example 3.1

The first and second stage amplifiers in Fig. 3.5 are replaced by two identical amplifiers each having $r_{in} = 100\ \text{k}\Omega$, $r_{out} = 100\ \Omega$, $K_v = 100$. The load is $r_L = 1\ \text{k}\Omega$. The whole is fed by a generator having a voltage of 1 mV when not loaded, and having a self resistance of $r_g = 20\ \text{k}\Omega$. Calculate v_{out_2} and the overall voltage amplification $\dfrac{v_{out_2}}{v_{in}}$.

Solution. The situation is the same as previously discussed except for the coupling between the two stages, where voltage coupling now takes place.

$$v_{out_2} = E_g \cdot \frac{r_{in_1}}{r_g + r_{in_1}} \cdot K_v \cdot \frac{r_{in_2}}{r_{out_1} + r_{in_2}} \cdot K_v \cdot \frac{r_L}{r_{out_2} + r_L}$$

That is

$$v_{out_2} = 10^{-3} \frac{100\ \text{k}\Omega}{20\ \text{k}\Omega + 100\ \text{k}\Omega} \cdot 100 \cdot \frac{100\ \text{k}\Omega}{100\ \text{k}\Omega + 100\ \text{k}\Omega} \cdot 100\ \frac{1\ \text{k}\Omega}{100\ \text{k}\Omega + 1\ \text{k}\Omega}$$

$$= 7.568\ \text{V}$$

To obtain v_{in} the input voltage coupling factor is used:

$$v_{in} = E_g \cdot \frac{r_{in_1}}{r_g + r_{in_1}} = 10^{-3} \cdot \frac{100\ \text{k}\Omega}{20\ \text{k}\Omega + 100\ \text{k}\Omega} = 0.83 \times 10^{-3}\ \text{V}$$

$$A_v = \text{Output voltage/input voltage}$$

$$= 7.568\ \text{V}/0.83 \times 10^{-3}\ \text{V}$$

$$= 9082$$

It may be noted that the output voltage is only 7568 times the signal generator voltage E_g of 1 mV. This is explained by the loading effect on the signal source of the input resistance of the first amplifier stage which causes the 1 mV signal to be reduced to 0.83 mV.

An important reason for ensuring near unity coupling factors is that by so doing, the dependence of the coupling factors on the input and output resistances is made small. For example, if in a voltage coupling factor $r_L/(r_{out_2} + r_L)$, the nominal values are $r_L = 1000\ \Omega$, $r_{out_2} = 100\ \Omega$ then a 10% change in either r_L or r_{out_2} causes a change of less than 1% in coupling factor. The overall amplification is thus relatively unaffected by any variation in any one resistance.

Check this yourself by calculation.

Frequency Response Effects

It was mentioned earlier that when alternating signals are applied to an amplifier phase-shifts occur and in general the gain is a complex quantity, having magnitude and phase. In fact, both magnitude and phase are functions of frequency. The *frequency-response curve* shown as a solid line in Fig. 3.6 exhibits *low-frequency* and *high-frequency* effects. The flat central region having greatest gain is the *mid-band region* and the gain here is called the *mid-band gain*. In this region the gain is

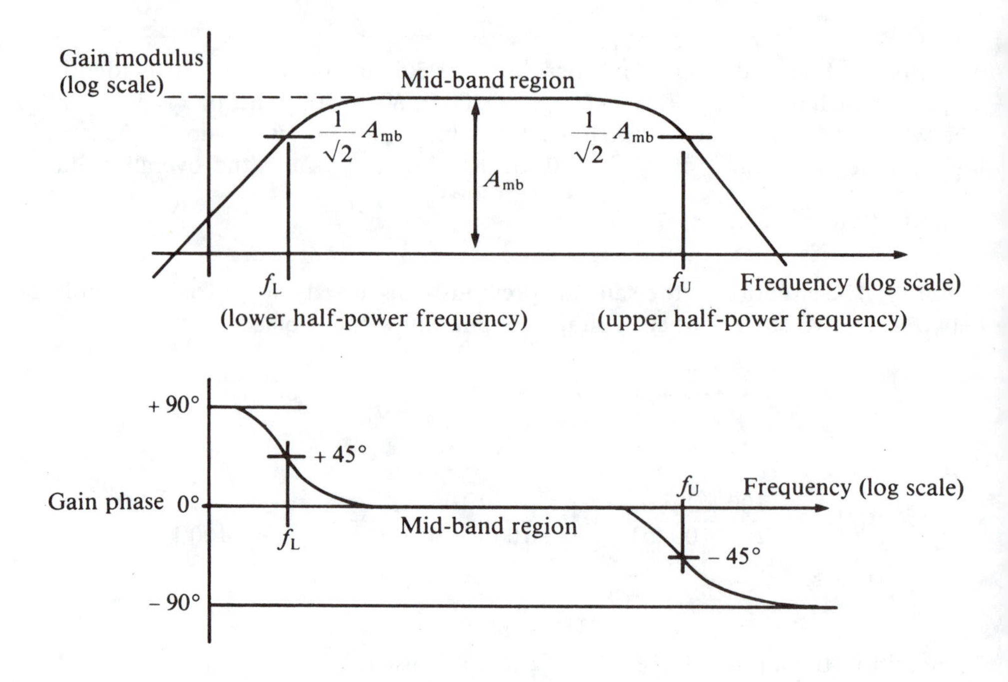

Fig. 3.6 Amplifier frequency-response curves illustrating low- and high-frequency effects.

largely unaffected by frequency-response effects and the gain modulus is relatively flat and the deviation of gain phase is very small. As shown in the figure, the mid-band phase is 0°; that is, the amplifier is non-inverting. The other case is that of 180° phase in the mid-band region, where the amplifier is of inverting type (that is, the input waveform is amplified and turned upside down). Inverting amplifiers are realized, for example, using an odd number of inverting amplifier stages of common-emitter type. In this discussion the amplifier is assumed to have 0° gain phase in the mid-band region. It is a simple matter to add the 180° phase difference to cover the inverting amplifier case, where necessary.

A gain which is negative indicates the signal is inverted not that it is attenuated. This would be indicated by a gain of magnitude less than unity.

Before a discussion of some of the causes of frequency-response effects in amplifiers is given, some more definitions are needed. The points where the gain modulus begins to fall off at the low and high-frequency ends of the mid-band region are important because they indicate the useful working range of the amplifier. These points are marked as f_L and f_U on Fig. 3.6, and are variously called the *cut-off frequencies*, the *corner frequencies*, or the *half-power frequencies*. Because the gain modulus falls off smoothly without sharp corners, some reference point has to be used to define the location of the edges of the mid-band region. The convention is to define the points f_L and f_U as occurring where the gain-modulus falls to $1/\sqrt{2}$ of the gain-modulus in the mid-band region. That is, if A(f) is the gain of the amplifier expressed as a complex function of signal frequency, then the lower half-power-point occurs when

The factor $1/\sqrt{2}$ is sometimes expressed as a percentage: 70.71%.

$$|A(f_L)| = \frac{1}{\sqrt{2}} |A(f), \text{mid-band}| \tag{3.22}$$

and the upper half-power-point occurs when

$$|A(f_U)| = \frac{1}{\sqrt{2}} |A(f), \text{ mid-band}| \tag{3.23}$$

At these frequencies for a constant input signal amplitude the output falls to $1/\sqrt{2}$ of its mid-band value. If the signal is a voltage and assuming the load is resistive, then, because the power delivered to the load is proportional to the voltage squared (it equals v_{out}^2/r_L), at the mid-band edges the power falls to $(1/\sqrt{2})^2$, that is to a half of the mid-band value. If the output signal is a current, the power also drops to a half at the band edge because power is proportional to current squared ($i_{out}^2 r_L$). This is why f_L and f_U are called the half-power frequencies. The distance between the two half-power frequencies, $f_U - f_L$ is called the *half-power bandwidth* or just *bandwidth* of the amplifier. In a wide-band amplifier f_U is much greater than f_L and so the bandwidth is approximately equal to f_U.

In dealing with the ratio of two signals, or the value of some signal relative to a reference level, it is common to use the *decibel* notation. Decibels are defined in terms of the logarithm of the ratio of the power levels of two signals (P_2/P_1), as follows:

$$\text{Power ratio} = 10 \log_{10} \frac{P_2}{P_1} \text{ decibels (dB)} \tag{3.24}$$

In many cases the signals are measured as voltages and because power is equal to voltage squared divided by the associated resistance:

$$\text{Power ratio} = 10 \log_{10} \frac{V_2^2}{R_2} \Big/ \frac{V_1^2}{R_1}$$

$$= 10 \log_{10} \frac{V_2^2}{V_1^2} \cdot \frac{R_1}{R_2}$$

$$= 20 \log_{10} \frac{V_2}{V_1} + 10 \log_{10} \frac{R_1}{R_2} \tag{3.25}$$

Since only ratios of the sinusoidal quantities V_1 and V_2 are used, it is not essential to use r.m.s. values and peak values may be used instead.

The first term on the right-hand side of Equation 3.25 is the voltage ratio expressed in decibels:

$$\text{Voltage ratio} = 20 \log_{10} \frac{V_2}{V_1} \text{ dB} \tag{3.26}$$

Also, power is equal to current squared multiplied by the associated load resistance. Equation 3.24 then becomes

$$\text{Power ratio} = 20 \log_{10} \frac{I_2}{I_1} + 10 \log_{10} \frac{R_2}{R_1} \tag{3.27}$$

r.m.s. or peak values may be used for I_1 and I_2.

where the first term on the right-hand side is the current ratio expressed in decibels. To obtain the decibel power ratio from the decibel current or voltage ratios, Equations 3.25 and 3.27 show that R_1 and R_2 would need to be known. Often, however, we are only interested in the ratio of current or voltage, as in A_V and A_I. In these cases it is not necessary to know R_1 and R_2 and it is convenient to express the ratio in decibels although it is not equal to the power gain in decibels unless $R_1 = R_2$.

Examples of ratios expressed as decibels are given in Table 3.1

Table 3.1

Current or voltage ratio	Decibel
1000	60
100	40
10	20
5	13.98 ≈ 14
3	6.02 ≈ 6
1	0
.7071 = $1/\sqrt{2}$	– 3.01 ≈ – 3
.1	–20
.01	–40

Exercise 3.4 Derive Equation 3.25. Also confirm the decibel values in Table 3.1.

One of the advantages of the decibel notation results from the basic properties of the logarithm of a product, $\log(A.B) = \log(A) + \log(B)$. This allows the gain of a chain of amplifiers or other linear circuits to be calculated easily by adding the decibel values. For example, an ideal amplifier of gain 100 (that is, 40 dB) in cascade with another of gain 10 (that is, 20 dB) provides an overall gain of 40 dB + 20 dB = 60 dB which corresponds to the decibel value of (100 × 10) = 1000 as can be confirmed from Table 3.1.

It can be seen from the table that the ratio 0.7071 corresponds to a decibel value of approximately –3 dB. For this reason the half-power points are sometimes called the 3 dB points because they are 3 dB down with respect to the decibel value of gain in the mid-band region.

H.W. Bode was a pioneer of feedback theory, his book *Network Analysis and Feedback Amplifier Design* (Krieger, 1975) is a classic but is not easy for the beginner.

It is common practice in plotting frequency-responses to use a vertical scale for gain modulus in decibels, and to use a horizontal axis with frequency plotted on a logarithmic scale. This is called a *Bode diagram*.

Gain changes in amplifiers at low frequencies can occur for a variety of reasons, but the most common cause is probably that of coupling capacitors. Fig. 3.7a shows a coupling capacitor C_1 inserted between two amplifying stages somewhere in an amplifier. Coupling capacitors are used to allow alternating signals to be coupled from the output of one amplifier stage to the input of another while at the same time

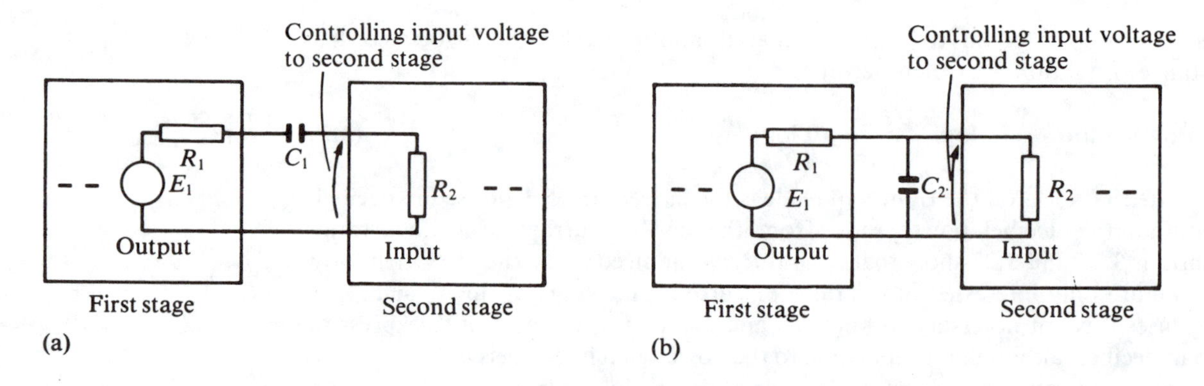

Fig. 3.7 How capacitance can cause (a) low-frequency and (b) high-frequency effects.

blocking constant voltages from being coupled. This is important, for example, when the collector output of a common-emitter BJT amplifier stage is coupled to the base-input of a following common-emitter BJT amplifying stage. Normally the collector output would be at a different quiescent voltage than the following base input, and a direct coupling with no capacitor would upset the quiescent condition of the circuit. Capacitors have infinite impedance at zero frequency, thus C_1 in Fig. 3.7a prevents any direct voltage from being coupled into the following stage. For alternating signals at a sufficiently high frequency the impedance of the coupling capacitor is low enough to allow good coupling. If the signal frequencies are lower than they should be then the impedance of the capacitor becomes significantly high and leads to a poor voltage coupling factor. A reduced overall gain ensues. The lower limit of usable signal frequencies can be found by analysing the circuit in Fig. 3.7a. The coupling capacitor C_1 may be considered to be in series with the output resistance of the first stage, R_1. Because of this the effective output impedance of the first stage is $R_1 + (1/j\omega C_1)$. Hence the voltage coupling factor (see Equation 3.18) is given by

Coupling capacitors are also to be found in FET circuits.

When alternating signals are used and circuits contain capacitances and inductances, then the presence of the reactances can be included by generalising the concept of resistance to that of *impedance*. This is explained in the appendix.

$$\text{Voltage coupling factor} = \frac{R_2}{R_2 + \left(R_1 + \dfrac{1}{j\omega C_1}\right)}$$

After rearranging, this becomes

$$\text{Voltage coupling factor} = \frac{R_2}{R_1 + R_2} \cdot \frac{1}{1 + \dfrac{1}{j\omega C_1(R_1 + R_2)}} \qquad (3.28)$$

This is the product of two factors. The first, $R_2/(R_1 + R_2)$, is the coupling factor, which occurs at frequencies high enough for the capacitor to be considered to act as a short circuit. The second is a frequency varying term which accounts for the presence of the coupling capacitor. Assuming for the moment that there are no other frequency response effects then the first term when multiplied by all the other coupling factors and gain constants must be equal to the amplifier gain at frequencies where C_1 has no effect. This is the mid-band gain A_{mb}. Thus the general expression for overall gain including the effect of the coupling capacitor is of the form

$$A = A_{mb} \cdot \frac{1}{1 + \dfrac{1}{j\omega C_1(R_1 + R_2)}} \qquad (3.29)$$

where A_{mb} is the mid-band gain obtained at frequencies which are high enough for the frequency response effect to be negligible.

There are three frequency regions of interest.

Signal Frequency Equal to the Half-Power Frequency ($\omega = \omega_L = 2\pi f_L$)

At this point the gain-modulus falls to $1/\sqrt{2}$ of its mid-band value. That is, the modulus of the right-hand factor of Equation 3.29 obeys the condition

$$\frac{1}{\sqrt{\left[1^2 + \left(\dfrac{1}{\omega_L C_1(R_1 + R_2)}\right)^2\right]}} = \frac{1}{\sqrt{2}}$$

Check the derivation of this yourself.

From this ω_L is easily found to give the half-power frequency.

$$\omega_L = \frac{1}{C_1(R_1+R_2)} \quad \text{or} \quad f_L = \frac{\omega_L}{2\pi} = \frac{1}{2\pi C_1(R_1+R_2)} \tag{3.30}$$

Using this result the equation for overall amplification (3.29) can be expressed in the more general form for a single low-frequency effect.

$$A = A_{mb}\cdot\frac{1}{1+\frac{\omega_L}{j\omega}} = A_{mb}\cdot\frac{1}{\sqrt{\left[1+\left(\frac{\omega_L}{\omega}\right)^2\right]}} \angle -\tan^{-1}\left(-\frac{\omega_L}{\omega}\right) \tag{3.31}$$

At $\omega = \omega_L$,

Check this yourself.

$$A = \frac{1}{\sqrt{2}} \angle +45° \tag{3.32}$$

This shows that at the half-power point the gain has a leading phase angle of $+45°$ as in Fig. 3.7.

Mid-band Frequencies ($\omega >> \omega_L$)

Using the general form (Equation 3.31) the overall gain at these frequencies becomes

$$A \approx A_{mb}$$

This is the expected result because at high frequencies the capacitor has negligible impedance and has no effect on the amplifier which therefore has a gain which is equal to the mid-band value.

Low Signal Frequencies ($\omega << \omega_L$)

In this case $\omega_L/\omega >> 1$ and the general form (Equation 3.31) becomes

$$A \approx A_{mb}\cdot\frac{1}{\frac{\omega_L}{j\omega}} = j\,\frac{A_{mb}}{\omega_L}\cdot\omega = A_{mb}\cdot\frac{\omega}{\omega_L}\angle +90°$$

This shows that as ω is reduced far below the half-power point the gain modulus falls linearly towards zero and the gain phase stays approximately constant at $+90°$.

The frequency response behaviour can also be obtained by direct computation of expression (3.29) over a range of values for ω, for some chosen. A_{mb}, C_1, R_1 and R_2. See Attikiouzel, J. *Pascal for Electronic Engineers* (Van Nostrand Reinhold (UK), 1984).

The frequency response behaviour of gain modulus and gain phase is shown in the low-frequency region of Fig. 3.6.

When designing an amplifier the coupling capacitor has to be chosen so that the cut-off frequency as given by Equation 3.30 is equal to, or less than, the specified value. This can be done by choosing a sufficiently large coupling capacitance. Larger capacitors tend to be more expensive, and where the designer has some control of R_1 and R_2 these can be chosen to be as large as conveniently possible to allow the required cut-off frequency to be achieved with the lowest value of coupling capacitor. Of course, in choosing R_1 and R_2 to be large, care must be taken to preserve the condition $R_2 >> R_1$ otherwise the voltage coupling factor, $R_1/(R_1+R_2)$, in the mid-band region suffers.

Consideration of costs again.

By means of appropriate active circuit techniques it is possible to avoid the use of coupling capacitors by ensuring that the quiescent voltages at the output and input of the successive amplifying stages are compatible. Such amplifiers are said to be *direct coupled*, and have the advantage that signal frequencies down to zero Hertz can be amplified, as shown by the broken line curve in Fig. 3.6. Direct coupled amplifiers are also used to save the cost of coupling capacitors. Although direct-coupled amplifiers usually require more active devices these are quite inexpensive, especially when the whole circuit is fabricated with integrated circuit technology, and there is often a net saving in cost.

Special transistor circuits have been developed which are suited to integrated circuit, see Ritchie, G.J., *Transistor Circuit Techniques*, Second edition, Van Nostrand Reinhold (International), 1987.

Having considered frequency-response effects at low frequencies, high-frequency effects are now examined. At high frequencies the most common cause of loss of amplification is stray capacitance appearing between a signal path and the common line. This is illustrated in Fig. 3.7b, where it is assumed that the capacitance C_2 appears at some internal point in the amplifier. Although shown as a single capacitor, in practice C_2 would often be the sum of several stray or parasitic capacitances. For example, it would include the output and input capacitances of active devices connected to that point and also the capacitance to the common line of the conductor which couples one amplifying stage to the next.

This circuit can also be analysed using the voltage coupling factor equation (3.18) as the effective load impedance is now given by C_2 in parallel with R_2. That is,

$$\text{Voltage coupling factor} = \frac{\dfrac{1}{j\omega C_2} \| R_2}{R_1 + \dfrac{1}{j\omega C_2} \| R_2}$$

From which,

As an exercise, work this out yourself.

$$\text{Voltage coupling factor} = \frac{R_2}{R_1 + R_2} \cdot \frac{1}{1 + j\omega C_2 \dfrac{R_1 R_2}{R_1 + R_2}} \tag{3.33}$$

The first term, $R_2/(R_1 + R_2)$, is again recognized as the coupling factor obtained when C_2 has no effect. The second term is frequency varying and represents the high-frequency effect due to C_2. The overall gain A, therefore, can be expressed in the form

$$A = A_{mb} \cdot \frac{1}{1 + j\omega C_2 \dfrac{R_1 R_2}{R_1 + R_2}} \tag{3.34}$$

Where A_{mb} is the gain in the mid-band region at frequencies which are not high enough to cause the frequency response effect to be exhibited.

The behaviour in the frequency regions of interest are:

(a) *Signal Frequencies Equal to the Half-Power Frequency* ($\omega = \omega_U = 2\pi f_U$). The half-power frequency occurs when the modulus of coupling factor falls to $1/\sqrt{2}$. From Equation 3.34, this happens when

The treatment here is similar to that for the low-frequency effect.

$$\frac{1}{\sqrt{\left[1 + \left(\omega_U C_2 \dfrac{R_1 R_2}{R_1 + R_2}\right)^2\right]}} = \frac{1}{\sqrt{2}}$$

From which

Do this yourself.

$$\omega_U = \frac{1}{C_2 \dfrac{R_1 R_2}{R_1 + R_2}} \quad \text{or} \quad f_U = \frac{\omega_U}{2\pi} = \frac{1}{2\pi C_2 \dfrac{R_1 R_2}{R_1 + R_2}} \tag{3.35}$$

This relationship can be used to re-express the overall gain (Equation 3.34) in the more general form

$$A = A_{mb} \cdot \frac{1}{1 + j\dfrac{\omega}{\omega_U}} \tag{3.36}$$

Note that when $\omega = \omega_U$

$$A = A_{mb} \cdot \frac{1}{1 + j} = \frac{A_{mb}}{\sqrt{2}} \cdot \underline{/-45°}$$

Hence, at the upper half-power frequency the gain phase lags that in the mid-band region by 45°.

(b) *Low Signal Frequencies* ($\omega << \omega_U$). The general relationship for overall gain (Equation 3.36) gives

$$A \approx A_{mb}$$

which is to be expected, since if the signal frequencies are low the impedance of C_2 is very high and has negligible effect on the amplifier gain.

Confirm this yourself.

(c) *High Signal Frequencies* ($\omega >> \omega_U$). In this case the general gain relationship (Equation 3.36) provides the following relationship:

$$A \approx A_{mb} \frac{1}{j\dfrac{\omega}{\omega_U}} = -j(\omega_U A_{mb}) \cdot \frac{1}{\omega} = (\omega_U A_{mb}) \frac{1}{\omega} \underline{/-90°} \tag{3.37}$$

This is an asymptotic condition; the phase approaches -90° and reaches it at the limit, $\omega \to \infty$. For the low-frequency effect the limit of -90° is achieved as $\omega \to 0$.

Hence as the signal frequency is increased the overall gain magnitude falls inversely with frequency while the gain phase stays approximately constant at -90.

The behaviour of the gain modulus and gain phase is shown at the high frequency end of Fig. 3.6. Equation 3.35 shows that to obtain a high upper half-power frequency, f_U, the designer must ensure that C_2 has a low value. This often means that careful attention has to be given to circuit layout so as to minimize stray capacitance. Equation 3.35 also shows that the frequency limit is increased by using low values of R_1 and R_2, but once again this must be done in a way which avoids a poor voltage coupling factor in the mid-band region; that is R_2 must not be small compared with R_1.

The above analysis of low-frequency and high-frequency effects has been carried out for the case of voltage coupling at the interior point in the amplifier. The analysis for the case of current coupling is similar and the relationships obtained for the upper and lower half-power frequencies are identical to those for the voltage coupling case (Equations 3.31 and 3.36). This is not altogether surprising considering the equivalence between the four amplifier representations (Fig. 3.4).

As an exercise, do this yourself.

Worked Example 3.2 An amplifier comprises a single stage of amplification. At the input of the amplifier a series capacitor C_1 is connected. The single stage has parameters $r_{in} = 10\ \text{k}\Omega$, $K_V = 100$, and $r_{out} = 1\ \text{k}\Omega$. The amplifier load is a resistance of value $R_L = 4\ \text{k}\Omega$. Stray capacitance C_2 appears across the output amplifier terminals and has a total effective value of 50 pF. Neglecting all other effects, calculate (i) the mid-band voltage gain, (ii) the upper half-power frequency, and (iii) the value of C_1 needed to provide a lower half-power frequency of 10 Hz.

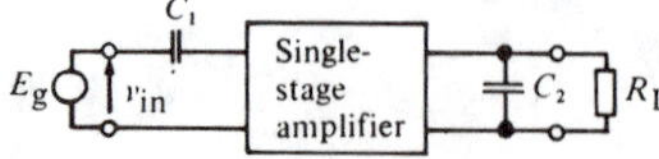

Solution: The overall amplification is obtained by multiplying together the input coupling factor, the gain constant K_V and the output coupling factor.

$$v_{out} = v_{in}\begin{pmatrix}\text{Input voltage}\\ \text{coupling factor}\end{pmatrix} \cdot K_V \cdot \begin{pmatrix}\text{Output voltage}\\ \text{coupling factor}\end{pmatrix}$$

$$= v_{in}\frac{r_{in}}{r_{in} + \dfrac{1}{j\omega C_1}} \cdot K_V \cdot \frac{R_L \| \dfrac{1}{j\omega C_2}}{r_{out} + R_L \| \dfrac{1}{j\omega C_2}}$$

which from the earlier discussion gives

$$v_{out} = v_{in} \cdot \frac{1}{1 + \dfrac{\omega_L}{j\omega}} \cdot K_V \cdot \frac{R_L}{r_{out} + R_L} \cdot \frac{1}{1 + j\dfrac{\omega}{\omega_U}}$$

where $\omega_L = 1/C_1 r_{in}$ and $\omega_U = 1/C(r_{out} \| R_L)$ are the lower and upper half-power points.

From this equation the parts of the question may be answered.

(i) Mid-band gain. In the mid-band region the high and low half-power frequencies are assumed to be widely separated and to have negligible effect on the gain in the mid-band region. This assumption is tested later.

Removing the frequency-variable terms in the last main equation gives

$$v_{out} = v_{in} K_V \cdot \frac{R_L}{r_{out} + R_L}$$

Hence,

$$A_v = \frac{v_{out}}{v_{in}} = 100 \times \frac{400}{100 + 400} = 80$$

(ii) For the upper half-power frequency,

$$f_U = \frac{\omega_U}{2\pi} = \frac{1}{2\pi}\frac{1}{C_1(r_{out} \| R_L)} = \frac{1}{2\pi \times 50 \times 10^{-12}\dfrac{1000 \times 4000}{1000 + 4000}}$$

Therefore,

$$f_U = 3.98 \text{ MHz}$$

This figure indicates that the upper and lower half-power frequencies are widely separated (by about six orders of magnitude) and so the assumption made in mid-band calculation is valid.

(iii) The coupling capacitance C_1 must satisfy

$$2\pi f_L = \omega_L = \frac{1}{C_1 r_{in}}$$

Hence

$$2\pi \times (10 \text{ Hz}) = \frac{1}{C_1 \times 10^4}$$

And so $C_1 = 1.59\ \mu F$.

High accuracy is not needed for C_1. The requirement is that C_1 should not be less than the above value. Capacitors (and resistors) are manufactured in a series of *preferred* values which are generally less expensive than special values. Consequently in practice a nearby preferred value above $1.59\ \mu F$ would be chosen for C_1, say $2.2\ \mu F$.

In Fig. 3.6 the frequency response (modulus and phase) is plotted versus frequency. Another way is to plot the gain modulus and phase for each frequency as a point on a phasor diagram in which the axes represent the real and imaginary parts of the gain. If this is done for all possible signal frequencies then the points join up to form a locus, which shows the variation of complex-gain with frequency. This is called a *Nyquist diagram*. The Nyquist diagram for the same amplifier as in Fig. 3.6 with a single low-frequency effect and a high-frequency effect is shown in Fig. 3.8a. It turns out that the separate loci for the low-frequency and high-frequency effects are each semicircular and the complete locus is therefore a circle. The locus is furthest away from the origin at mid-band frequencies where the gain modulus is greatest. At the frequency limits $\omega \rightarrow \infty$ and $\omega \rightarrow 0$ then the locus approaches the origin of the axes indicating zero gain modulus as expected.

Nyquist is another of the pioneers of feedback theory.

Show that the positions of the points ω_U and ω_L on the Nyquist diagram are correctly marked.

In practice the frequency response behaviour of an amplifier can be a good deal more complicated than that described. Causes other than coupling capacitors and stray capacitances give rise to frequency-response effects. For example, if transformers are used to couple signals between stages, a low-frequency effect is introduced because of the inability of transformers to work down to zero frequency. In addition, the self-inductance of conductors can have a significant impedance at higher frequencies. There is also the fundamental upper frequency limit which exists in any active device owing to the finite time it takes for the output of a device to respond to an instantaneous change in input signal.

Further treatment of frequency response effects is to be found in Ahmed, H., and Spreadbury, P.J. *Electronics for Engineers* (Cambridge University Press, 1978) and also Millman, J. *Microelectronics* (McGraw-Hill, 1979).

A second reason for more complicated behaviour is that usually more than one frequency effect is present. For example, a two-stage amplifier could exhibit three

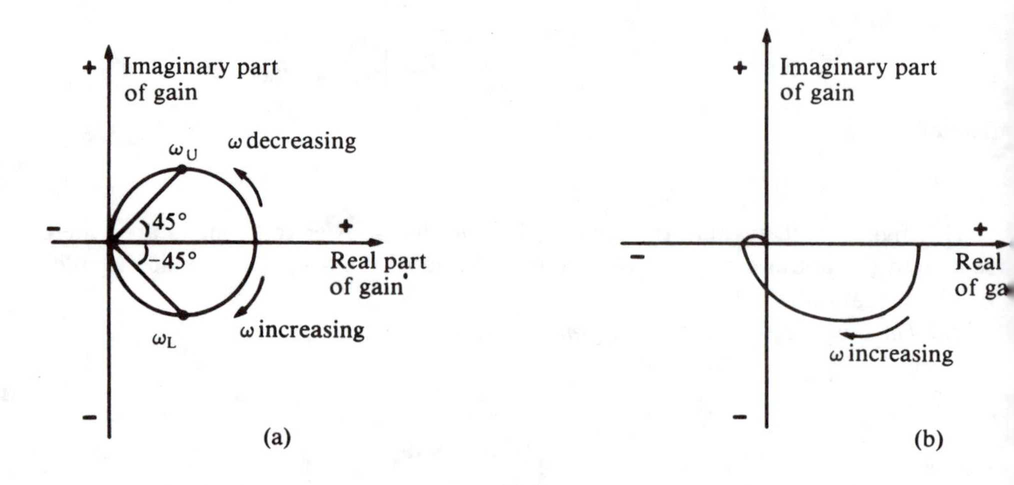

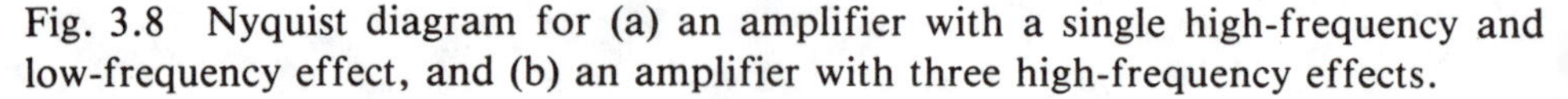
Fig. 3.8 Nyquist diagram for (a) an amplifier with a single high-frequency and low-frequency effect, and (b) an amplifier with three high-frequency effects.

(or more) high-frequency effects owing to shunt stray capacitances; one between input signal and first stage, one between first and second stage, and one between second stage and the load. Assume in these cases of multiple effects the frequency-dependent factors are multiplied together. Suppose for simplicity that each of the three coupling factors has associated with it the same cut off frequency ω_1; then the overall gain will be of the form

$$A = A_{mb} \cdot \frac{1}{1 + j\frac{\omega}{\omega_1}} \cdot \frac{1}{1 + j\frac{\omega}{\omega_1}} \cdot \frac{1}{1 + j\frac{\omega}{\omega_1}} \qquad (3.38)$$

The multiplication of these factors is to assume that they do not interact in the circuit. This is usually, but not always, a valid assumption. An example of interacting effects is the coupling and emitter by-pass capacitances in the common-emitter amplifier, see Millman, J. *Microelectronics* (McGraw-Hill, 1979), pp. 455 et seq.

With increasing frequency the gain modulus of A now falls to zero more rapidly than for the case of a single frequency effect. Also, the limiting phase shift at high frequencies with the three effects is greater, given by $3 \times \angle{-90^\circ} = \angle{-270^\circ}$. The Nyquist diagram in this case is shown in Fig. 3.8b. This illustrates a point which, in general, applies to any amplifier where multiple frequency effects occur. At some frequency the additional phase shift introduced because of the frequency-response effects may equal or exceed 180°. The significance of this is that an amplifier which has been designed to have negative feedback in the mid-band region can at some frequencies operate in positive feedback mode. This raises the question of possible instability, which is taken up in Chapter 5.

Another cause of more complicated frequency-response behaviour is that inductive and capacitive components can sometimes appear at other places in the circuit than those indicated in Fig. 3.7. For example, the parasitic base-collector capacitor C_{bc} of a transistor appears between input and output when the transistor is used in common-emitter configuration. General circuit analysis methods, for example nodal or mesh analysis, can always be used to analyse the circuits in these cases. If the circuit is complicated, a hand solution can be lengthy and computer simulation techniques may be preferred.

An introductory text on this topic is Fidler, J.K. and Nightingale, C. *Computer Aided Circuit Design* (Nelson, 1978).

Differential Amplifiers

Any amplifier has two input terminals and two output terminals. In many of the commonly used amplifier circuits one of the input terminals and one of the output terminals are connected to a *common-rail* which is often the zero volts line. An example is the common-emitter transistor amplifier circuit, whose name indicates that the transistor emitter is common to both the input and output terminals. This condition can be shown on the amplifier model as a conductor connecting an input to an output terminal. Fig. 3.9 shows this for a voltage amplifier. Because one of the

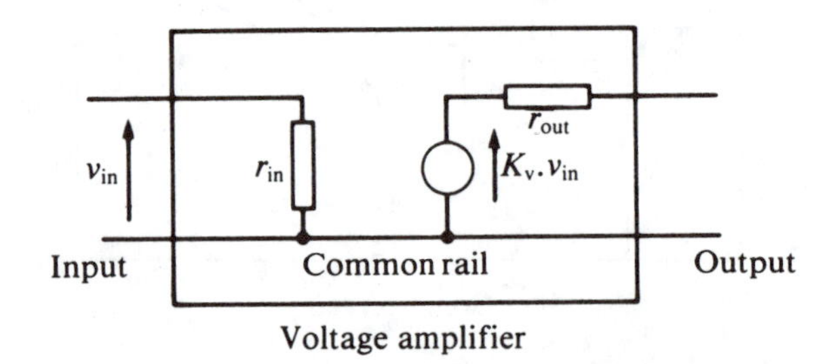

Fig. 3.9 Single-ended voltage amplifier.

input terminals is connected to the common rail, this means that the amplifier has one active input terminal and not two. For this reason the amplifier is said to be a *single-ended-input* amplifier. Similarly at the amplifier output because one terminal is connected to the common-rail the amplifier is said to have a *single-ended output*.

Amplifier circuits having single-ended inputs and outputs are useful and satisfy many applications. However, amplifiers in which the input signal is not restricted to having one end at common rail voltage are more versatile. A voltage amplifier of this type is called a *differential amplifier*.

Do not confuse this with a *differentiator* circuit whose output is proportional to the time derivative of the input signal (see Chapter 6).

Suppose such an amplifier is driven by two signals having voltages V_1 and V_2 measured with respect to the common rail, as shown in Fig. 3.10. Only the voltage difference across the input terminals of the amplifier is amplified. Hence the difference voltage $(V_1 - V_2)$ appears amplified at the amplifier output,

$$V_{out} = A_{dm}(V_1 - V_2) \tag{3.39}$$

It is usual to call $(V_1 - V_2)$ the *differential-mode voltage*, V_{dm}, and the gain constant, A_{dm}, is called the *differential-mode amplification*, or simply the *differential gain*. Thus,

$$V_{out} = A_{dm} \cdot V_{dm} \tag{3.40}$$

The differential amplifier can be operated as a single-ended-input amplifier. This is done by setting either input at the common rail voltage. Choosing first the lower input terminal, then this can be made to be at the same voltage as the common rail by setting $V_2 = 0$. Equation 3.39 gives the output voltage to be

$$V_{out} = A_{dm} V_1 \tag{3.41}$$

which shows that provided A_{dm} is assumed to be positive the output voltage is the same polarity as the upper terminal input voltage V_1. This terminal is called the non-inverting input terminal and labelled with a plus sign.

Choosing the upper terminal to be connected to the common rail, and setting $V_1 = 0$, Equation 3.39 now gives

$$V_{out} = -A_{dm} \cdot V_2 \tag{3.42}$$

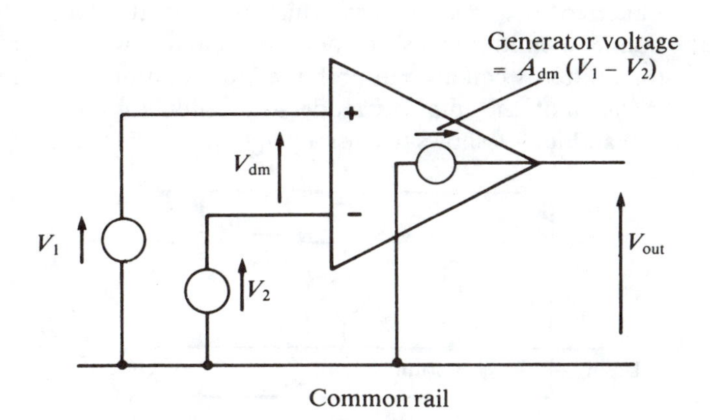

Fig. 3.10 Differential amplifier.

This shows that the output voltage is of opposite polarity to that of the lower terminal input voltage V_2. This terminal is therefore called the inverting input terminal and labelled with a minus sign.

Note: The plus and minus signs refer to the gain polarities and not to the signal polarities which in general can be any polarity.

One important application of differential amplifiers is to the amplification of signals which of necessity are generated some distance from the point where the signals are processed. A typical example is that of a temperature transducer located inside a machine to measure hot spots within the machine. The simple approach is to connect one terminal of the transducer to a common point or *earth* near the machine and run a wire from the other terminal of the transducer to a single-ended-input amplifier. However, in many cases this approach is not workable. One reason is that the wire carrying the signal acts as a receiving aerial for nearby electrical interference. The signal appearing at the amplifier is therefore equal to the wanted signal v_s plus an unwanted noise signal v_n owing to the interference. Screening can lessen the noise signal but there can still be problems from this source, especially if the transducer signal is small, as is often the case. Also, it is quite likely that the common point or earth near the transducer is not at exactly the same potential as the common rail in the amplifier if they are some distance apart (since there are always fluctuating voltages between any two remote earth points caused by other electrical phenomena). This difference in common point voltage, v_c say, is treated by a single-ended-input amplifier as part of the signal and is amplified as well. These problems are substantially overcome by running a second wire from the other terminal of the transducer, and both wires are now taken to the input of a differential input amplifier. The arrangement is shown schematically in Fig. 3.11. An essential requirement is that both signal wires run very closely together so that whatever noise voltage is picked up is the same in both signal wires. To derive an equation for the amplifier output, the first step is to find the voltages V_1 and V_2 appearing at the amplifier input and then Equation 3.39 can be used to obtain v_{out}. The voltages V_1 and V_2 are measured with respect to the amplifier common-rail voltage. Taking this as the voltage reference point then a loop can be followed via v_c, v_s and v_n on the upper signal line, to the point where V_1 is measured. This gives

Screening is a subject in its own right, an authoritative text is Morrison, R. *Grounding and Shielding Techniques in Instrumentation*, 2nd ed. (Wiley, 1977).

One way to ensure two wires run closely together is to twist them.

$$V_1 = v_c + v_s + v_n \tag{3.43}$$

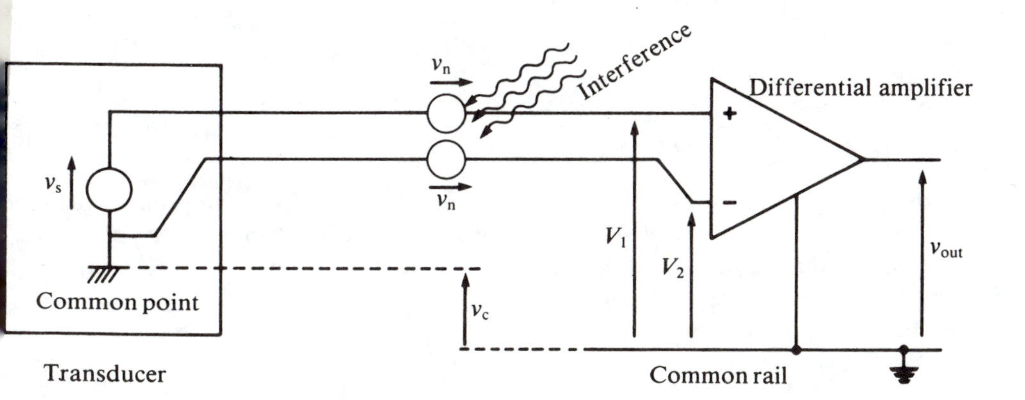

Fig. 3.11 Reduction of interference using a differential amplifier.

Similarly a loop can be followed via v_c and v_n on the lower signal wire to give

$$V_2 = v_c + v_n \tag{3.44}$$

Using the equation for amplifier output voltage (3.39),

$$v_{out} = A_{dm}(V_1 - V_2) = A_{dm}((v_c + v_s + v_n) - (v_c + v_n))$$

Hence

$$v_{out} = A_{dm} \cdot v_s \tag{3.45}$$

which shows that only the wanted component v_s appears at the amplifier output. This rejection of unwanted signals is very significant.

This technique can succeed even where the unwanted component greatly exceeds the wanted signal (see Worked Example 3.3).

Differential amplifiers are widely used in electronic instrumentation applications, and are often called *instrumentation amplifiers*. Another application of the differential amplifier is in the construction of a differential oscilloscope. Such an oscilloscope is able to indicate the voltage waveform across a component in a test circuit, neither end of which is at zero volts.

With this, and other applications of differential amplifiers, ideally only the difference of the voltages at the amplifier input terminals should control the amplifier output. With practical amplifiers this is not exactly the case. Equations 3.41 and 3.42 show that the amplification from the non-inverting input terminal should be equal in magnitude to that from the inverting input terminal. However, because there are two signal paths through any differential amplifier, for at least part of the internal circuitry the two signals must flow through different devices and therefore it is inevitable that small differences exist between the amplifications measured from each input terminal. Hence Equations 3.41 and 3.42 are of the form

Gain from non-inverting input, $v_{out} = A_1 v_1$ when $v_2 = 0$

Gain from inverting input, $v_{out} = -A_2 v_2$ when $v_2 = 0$

In practice A_1 and A_2 are approximately equal to A_{dm} but not exactly so.

The equation for output voltage (3.39) now becomes

$$v_{out} = A_1 V_1 - A_2 V_2$$

Splitting this up

$$v_{out} = \left(\frac{A_1}{2} + \frac{A_1}{2}\right) V_1 - \left(\frac{A_2}{2} + \frac{A_2}{2}\right) V_2$$

and regrouping gives

$$v_{out} = \frac{A_1 + A_2}{2} \cdot (V_1 - V_2) + (A_1 - A_2)\frac{V_1 + V_2}{2} \tag{3.46}$$

With the equation in this form the desired and undesired components can be identified. The factor $(V_1 - V_2)$ is recognized as the differential mode voltage v_{dm}. The factor which multiplies it must, therefore, be the differential mode gain, which is equal to the average of A_1 and A_2,

$$A_{dm} = \frac{A_1 + A_2}{2}$$

The right-hand term in Equation 3.46 is seen to arise because $(A_1 - A_2)$ is non-zero, that is because $A_1 \neq A_2$. The factor $\frac{1}{2}(V_1 + V_2)$ in this term is the average of the input terminal voltages measured with respect to the common point, and is therefore called the *common-mode voltage* v_{cm},

$$v_{cm} = \frac{V_1 + V_2}{2}$$

The factor which multiplies v_{cm} in Equation 3.46 represents the unwanted appearance of a term owing to v_{cm} at the output. This factor is called the *common-mode amplification* A_{cm}, or simply the *common-mode gain*

$$A_{cm} = A_1 - A_2$$

The practical measurement of A_{cm} using this equation is not successful because A_1 and A_2 are close in value. See problem 3.11 for a better way.

Using these definitions Equation 3.46 for output voltage becomes

$$v_{out} = A_{dm} \cdot v_{dm} + A_{cm} \cdot v_{cm} \tag{3.47}$$

Exercise 3.5

Derive the following equations, and verify the equivalent circuit for v_{cm} and v_{dm} shown in the margin.

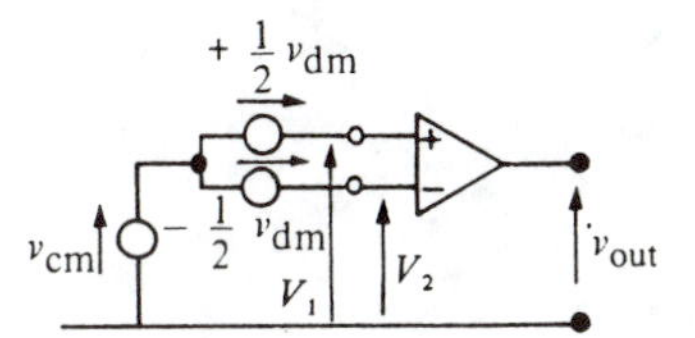

$$A_1 = A_{dm} + \frac{1}{2} A_{cm} \qquad A_2 = A_{dm} - \frac{1}{2} A_{cm}$$

$$V_1 = v_{cm} + \frac{1}{2} v_{dm} \qquad V_2 = v_{cm} - \frac{1}{2} v_{dm}$$

Considering the analysis of the instrumentation amplifier again (Fig. 3.11), Equation 3.45 shows that the signal v_s is equal to the differential-mode voltage v_{dm} and is thus amplified by A_{dm}. The common-mode voltage is given by

$$v_{cm} = \frac{V_1 + V_2}{2} = \frac{(v_c + v_s + v_n) + (v_c + v_n)}{2}$$

$$= v_c + v_n + \frac{1}{2} v_s$$

The common-mode voltage thus contains the unwanted components v_c and v_n. In many cases $v_c + v_n$ is large compared to the signal owing, for example, to the induction of 50 Hz signals for mains powered equipment. In other words, in practice the common-mode signal is often larger than the differential-mode signal.

It is therefore desirable to have the common mode amplification at a low value to keep the unwanted output component owing to common-mode inputs at an acceptable value. A measure of how good a differential amplifier is at not responding to common-mode inputs is the *Common-Mode Rejection Ratio* (CMRR) which is defined as

$$\text{CMRR} = \left| \frac{A_{dm}}{A_{cm}} \right| \tag{3.48}$$

Worked Example 3.3

A transducer which generates a 50 mV signal is connected to a differential amplifier by two signal wires which pick up unwanted common-mode voltage of 5 V. The amplifier output due to the transducer signal is required to be 10 V. The unwanted output component owing to the common-mode input is to be less than 1% of the wanted component. Specify the amplifier differential-mode gain and common-mode rejection ratio.

Solution: From Equation 3.47 the wanted output component is

$$v_{out} = A_{dm} \cdot v_{dm}$$

Substituting values then

$$10\ \text{V} = A_{dm} \cdot 50\ \text{mV}$$

Hence, the differential mode gain is to be specified at

$$A_{dm} = \frac{10\ \text{V}}{50\ \text{mV}} = 200$$

Equation 3.55 also gives the unwanted output component

$$v_{out} = A_{cm} \cdot v_{cm}$$

That is, (1% of 10 V) = $A_{cm} \cdot 5$ V. Hence

$$A_{cm} = \frac{1\%\ \text{of}\ 10\ \text{V}}{5\ \text{V}} = \frac{0.1\ \text{V}}{5\ \text{V}} = 0.02$$

Hence, the specified CMRR is

$$\text{CMRR} \geqslant \frac{200}{0.02} = 10^4$$

A CMRR of 80 dB is easily obtained with modern integrated differential amplifiers.

It is common practice to express CMRR in decibels, 20 $\log_{10}$ (CMRR) dB, and so a minimum CMRR of 80 dB is required for this application.

Summary

An ideal amplifier can be either current controlled or voltage controlled at its input, while its output can be provided by an ideal current generator or voltage generator. Thus there are four types of ideal amplifier: current-controlled current-source, current-controlled voltage-source, voltage-controlled voltage-source, and voltage-controlled current-source.

Real amplifiers can be modelled by attaching *input* and *output resistances* to a controlled source. In general any type of controlled source can be chosen for this because the four possible representations are equivalent and when used to analyse amplifier behaviour will provide identical results. For convenience the model can be chosen by considering input and output resistance values. For example, if r_{in} is high a controlled source which is voltage controlled may be chosen, and if r_{out} is low a controlled source with a voltage generator may be chosen.

A complete amplifier system consists of a signal source having a self resistance, an amplifier, and a load resistance. The amplifier could comprise a number of separate amplifier stages connected in cascade, and each amplifying stage has its own r_{in} and r_{out}. To analyse such a system then the individual stage gain and the *coupling factors* are multiplied together. Coupling factors are of *voltage* type or *current* type and account for the loss of signal due to loading effects at points in the amplifier where a signal source is coupled to amplifying stage input or to the load. For good coupling of voltage signals, the effective source resistance at the point of coupling should be much less than the effective load resistance. For good current coupling the opposite condition should apply.

As signal frequencies are increased the presence of *high-frequency effects* eventually cause the gain to reduce in magnitude towards zero. The main reason for this is usually stray capacitance which can occur for example because of the layout of the circuit. The gain may also drop at low frequencies owing usually to coupling capacitors. In *direct coupled* amplifiers the use of coupling capacitors is avoided and so useful amplification extends down to signal frequencies to 0 Hz (i.e. d.c.). The variations with frequency of gain modulus and phase due to these effects can be calculated, as can the upper and lower *half-power frequencies* which mark the onset of the high- and low-frequency effects. The *bandwidth* of the amplifier is the difference of the two half-power frequencies.

Many amplifiers are *single-ended* in that they have one input and one output terminal connected to a common rail. The *differential amplifier* has both input terminals free from the common rail and can be used to amplify the difference of two voltage signals. This difference voltage is called the *differential-mode voltage*, v_{dm} and the amount by which it is amplified is called the *differential-mode gain*, A_{dm}. One important application of differential amplifiers is to the amplification of signals in the presence of interference. Used in this application the amplifier is called an *instrumentation amplifier*. The interference constitutes a *common-mode voltage* v_{cm} which is the average of the two input voltages and it is important that the associated *common-mode amplification* A_{cm} is low. A measure of the quality of a differential-mode amplifier is the *common-mode rejection ratio*, A_{dm}/A_{cm}.

Important though this application is, the greatest significance of differential amplifiers arises because they are cheaply made and as is seen later are convenient to use with feedback since the polarity can be chosen by connecting to the appropriate input terminal. Called *op. amps.*, they enable circuits to be designed using feedback techniques to provide a wide range signal processing behaviour with very good performance. This is of major importance and Chapters 6 onwards of this text are devoted to the subject.

Before this, though, more of the circuit and other properties of feedback amplifiers are developed.

This chapter has shown that amplifiers take on different representations. They can be considered to be controlled by a voltage or current. Their outputs can also be considered as voltages or currents. Feedback can therefore be applied in different ways and a number of basic feedback circuits exist. This is the subject of the next chapter.

Problems

3.1 An amplifier is made from a cascade of four ideal amplifier stages with the output of one ideal amplifier connected to the input of the next. The first stage is a c.c.c.s. and the following stages are a c.c.v.s., a v.c.v.s., and a v.c.c.s., in that order. The respective gain constants are –50, 800 Ω, 40, and -30×10^{-2} S. If a 1 μA signal is applied to the first stage what current and voltage appears across a 1 kΩ resistive load at the final output?

3.2 Show that if no load is connected across the output terminals of a voltage amplifier which has finite r_{in} and r_{out}, the measured voltage gain A_V is equal to the amplifier gain constant K_V.

3.3 A real amplifier is modelled using a voltage-controlled voltage source, $K_V = 100$, and resistances $r_{in} = 10\ k\Omega$ and $r_{out} = 5\ k\Omega$. Derive equivalent representations of the amplifier using the other controlled sources.

3.4 If an amplifier has an ideal input resistance $r_{in} = 0$ but a finite output resistance, only two of the representations of Fig. 3.4 exist. Why is this, and which two are they?

3.5 An amplifier comprises two voltage amplifier stages in cascade. The first amplifier has r_{in}, r_{out} and K_V of 100 kΩ, 1 kΩ, and +50 respectively. The same parameters for the second stage are 200 kΩ, 100 Ω, –60, respectively. The amplifier load is a resistor of 10 kΩ. The applied signal is a 2 mV voltage generator with a self resistance of $R_g = 20\ k\Omega$. Calculate v_{out}, i_{out}, A_V and A_I.

3.6 Change the order of the first and second stages in the previous question and again calculate v_{out}, i_{out}, A_V and A_I. Note that changing the order has altered the amplifier behaviour. Why is this?

3.7 A real amplifier has $r_{in} = 10\ k\Omega$, $K_V = 100$, and $r_{out} = 2\ k\Omega$. The load is $R_L = 8\ k\Omega$ across which is a small stray capacitance C_2. The amplifier output terminal is connected to the load by a coupling capacitor C_1. Calculate the minimum value of C_1 and the maximum permitted value of C_2 if the bandwidth of the amplifier voltage gain is to extend from 20 Hz to 5 MHz. What value is the mid-band gain?

3.8 An amplifier has a mid-band gain of –100 and a lower half-power frequency at 100 Hz owing to a single low-frequency effect. Calculate the gain magnitude and phase when the signal frequency is (i) twice the half-power frequency and (ii) half the half-power frequency.

3.9 An amplifier has two high-frequency effects caused by stray capacitances at separate points in the circuit. Each effect contributes a term $1/(1 + j\omega/\omega_2)$ (see Equation 3.36) to the expression for gain. Because there are two effects the upper half-power frequency is not now equal to ω_2. Derive a formula for the upper half-power frequency and also show that the gain phase angle is not –45° as is the case for a single frequency-response effect.

3.10 An amplifier has a complex gain given by

$$\hat{A} = 100/(1 + j\frac{\omega}{\omega_1})(1 + j\frac{\omega}{\omega_1})$$

where $\omega_1 = 2\pi \times 10^7$ rad/s. Calculate the frequencies at which the Nyquist diagram

(i) intersects the negative imaginary axis,
(ii) intersects the negative real axis.

3.11 A voltage V measured with respect to the common rail is applied to one input terminal of a differential amplifier. The two input terminals are shorted together thus making $V_1 = V_2$. Show that Equation 3.47 for output voltage now becomes

$$v_{out} = A_{dm} \cdot V$$

(This provides a useful practical way to measure A_{dm}.)

3.12 Inputs of $V_1 = 1.01$ V and $V_2 = 0.99$ V are applied to a differential amplifier and cause an output voltage of 10 V. The input voltages are each reduced by 5 V and the amplifier output is observed to change to 10.2 V. Calculate A_{dm}, A_{cm} and CMRR.

Feedback Amplifier Circuits 4

Objectives

☐ To explain that feedback can be applied to an amplifier in four ways.
☐ To distinguish between parallel and series feedback.
☐ To quantify the effect that feedback has on amplifier input and output impedances.
☐ To show that large amounts of applied negative feedback cause the amplifier to behave as one of the four basic types of amplifier.
☐ To describe some typical feedback amplifier circuits and analyse their behaviour.

The Four Feedback Circuit Configurations

To form a feedback circuit three functional blocks are required:

(i) a forward amplifier,
(ii) a feedback block which samples the output of the forward amplifier, and
(iii) a mixer which subtracts the feedback signal from the main incoming signal and inputs the difference to the forward amplifier (see Fig. 2.1).

The functional blocks can be configured in four basic ways (see Fig. 4.1) providing feedback circuits which have different but individually desirable properties. The following describes the feedback circuits and the different properties of the feedback amplifier circuits.

Consider the output signal of the forward amplifier. In general, as was shown in the previous chapter, an amplifier supplies both signal current *and* signal voltage to the load, therefore, the choice exists for the feedback block β either to sample output voltage or current. To sample the output voltage the feedback block must act rather like a voltmeter and consequently is connected across the output terminals of the forward amplifier. Because it is connected *across* the output this is called *parallel-output* sampling or *voltage feedback*. To sample the output current the feedback block must act rather like an ammeter and so is connected in series with the loop which is formed by the amplifier output and the load. This is called *series-output* sampling or *current feedback*. Arrangements (b) and (c) of Fig. 4.1 have parallel-output sampling and the other two (a and d) have series-output sampling.

Now consider the input signal to the forward amplifier. This can be considered as a voltage or a current. The mixer which provides this signal can have two forms. In one form, the mixer takes the main incoming signal voltage and subtracts from it the feedback signal considered as a voltage source. A simple way to subtract one voltage from another is to series connect them in opposing direction. This is shown in arrangements (c) and (d) of Fig. 4.1 and is referred to as *series-input* feedback. In the other form, the mixer subtracts the feedback signal considered as a current source, from the main incoming signal current. A simple way to subtract one current source from another is to connect the two in parallel as shown in (a) and (b) of Fig. 4.1. Note that the direction of the feedback signal current is shown coming out

The concept of duality runs through much of electrical circuit theory. Properties which apply to voltages often have counter parts which apply to currents.

of the top node to conform with the convention that the mixer subtracts the feedback signal from the main incoming signal. This type of mixing is called *shunt-input* feedback.

Different names for the same thing. It can often happen that an unfamiliar name represents something we are already familiar with.

Thus there are the four types of configuration shown in the figure and called *shunt current*, *shunt voltage*, *series voltage* and *series current*, respectively. Other names are used for these types of input and output connections. Examples are *voltage output*, *node sampling* for parallel output; *loop sampling* for series output; *voltage input*, *loop input* for series input and *current input* and *node input* for parallel input.

Now consider the form of the general feedback relationship (Equation 2.5) for the four configurations. Taking the shunt voltage parallel arrangement of Fig. 4.1b, the expression for closed-loop amplification is obtained as follows:

The input and output signals for the forward amplifier are the input current i_{ia} and output voltage v_o. Therefore the transfer parameter of interest is the transresistance R_T of the amplifier,

$$v_o = R_T i_{ia} \tag{4.1}$$

The output voltage of the amplifier is sampled by the feedback block and a proportion is returned to the mixer as a current, thus

$$i_f = \beta_G . v_o \tag{4.2}$$

The suffix G has been added because the feedback fraction in this case has units of amperes per volt; that is, Siemens, the units of conductance.

Formally this is Kirchhoff's Current Law.

The mixer constrains the currents to obey the relationship

$$i_{ia} = i_i - i_f \tag{4.3}$$

We want to keep the input/output quantities i_i, v_o, and eliminate the internal quantity i_f.

It is now a simple matter to combine these equations to obtain a single equation relating input i_i to output v_o. Substituting Equation 4.3 into Equation 4.1,

$$v_o = R_T(i_i - i_f)$$

and using Equation 4.2 gives

$$v_o = R_T(i_i - \beta_G . v_o)$$

Rearranging this expression gives the desired result

$$\text{Closed-loop gain} = \frac{v_o}{i_i} = \frac{R_T}{1 + R_T\beta_G}\ \Omega \tag{4.4}$$

The other configurations are analysed in a similar way and the results are shown in Table 4.1.

Exercise 4.1 Verify the contents of Table 4.1.

It is worth noting that each of the expressions for closed-loop gain conforms to the general form of open-loop gain divided by $(1 + A\beta)$. The open-loop transfer parameter is of the same type as the closed-loop transfer parameter. Thus, for example, in the shunt voltage case the open-loop gain is R_T and the closed-loop gain after feedback is applied is R_{Tf}, are both transresistances. It follows that the dividing factor $(1 + A\beta)$ is dimensionless, as are the $A\beta$ products: $A_I\beta_I$, $R_T\beta_G$, $G_T\beta_R$, A_V,β_V. Two of the $A\beta$ products contain transfer parameters which are not dimensionless,

Table 4.1 Feedback Signals and Closed-Loop Gains

Type of feedback	Input quantities X_i, X_{ia}, X_f	Output quantities	Forward amplification A	Feedback fraction β	Closed-loop gain A_f
Shunt current	currents i_i, i_{ia}, i_f	current i_o	$A_I = \frac{i_o}{i_{ia}}$	$\beta_I = \frac{i_f}{i_o}$	$A_{If} = \frac{A_I}{1 + A_I \cdot \beta_I}$
Shunt voltage	currents i_i, i_{ia}, i_f	voltage v_o	$R_T = \frac{v_o}{i_{ia}}$	$\beta_G = \frac{i_f}{v_o}$ S	$R_{Tf} = \frac{R_T}{1 + R_T\beta_G}$ Ω
Series voltage	voltages v_i, v_{ia}, v_f	voltage v_o	$A_V = \frac{v_o}{v_{ia}}$	$\beta_V = \frac{v_f}{v_o}$	$A_{Vf} = \frac{A_V}{1 + A_V \cdot \beta_V}$
Series current	voltages v_i, v_{ia}, v_f	current i_o	$G_T = \frac{i_o}{v_{ia}}$	$\beta_R = \frac{v_f}{i_o}$ Ω	$G_{Tf} = \frac{G_T}{1 + G_T \cdot \beta_R}$ S

In general amplification and feedback fraction may have phase shift and so the quantities A_I, R_T, A_V, G_T, β_I, β_G, β_V, and β_R are complex quantities.

Notice the symmetry that is present in the table. Any entry can be covered up and correctly guessed by looking at other entries in the table.

R_T (ohms) and G_T (siemens) and so the multiplying feedback fractions, β_G and β_R respectively must have reciprocal dimensions (siemens and ohms).

Because the closed-loop expression conforms to the general form, the benefits of negative feedback described in Chapter 2 apply to each of the four feedback configurations shown in Fig. 4.1. The effect of increasing the amount of applied negative feedback is to reduce the sensitivity of the gain of the feedback circuit to variations in the forward gain A_I, A_v, R_T, or G_T. In the limit as the loop-gain approaches infinity then the gain after feedback equals the reciprocal of the feedback fraction β and is entirely independent of the gain of the forward amplifier. In this respect the effect of negative feedback is to improve whatever type of amplifier is used in the forward gain path so that it more closely approximates one of the ideal amplifiers, or controlled sources, shown in Fig. 3.2, the type being determined by the feedback type. Figs. 3.2 and 4.1 have been drawn so that the various controlled sources and the feedback configurations correspond with each other. Thus, shunt current feedback (Fig. 4.1a) improves the forward amplifier so that it more closely approximates a current-controlled current source, or ideal current amplifier (Fig. 3.2a). For this statement to be completely true it would be necessary for negative feedback to improve the amplifier input and output impedances as well as the gain. This is indeed the case and is shown to be so in the next section.

The general form for closed loop gain is

$$A_f = \frac{A}{1 + A\beta}$$

Note that $1/\beta$ may be dimensionless or may have units of resistance or conductance depending on the feedback configuration.

Effect of Feedback on Input and Output Impedance

In some applications it can be at least as important for an amplifier to have good input or output impedances as it is to have amplification. As was seen in the previous chapter, voltage or current coupling factors close to unity provide protection against variations in source and load impedance which occur, for example, because of frequency effects. In terms of amplifier parameters this means that for input voltage signals a high input impedance is desirable, while for input current signals a low input impedance is desirable. Conversely, for amplifier outputs which are voltage sources a low output impedance is desirable and for outputs which are current sources a high output impedance is desirable.

Remember the word *impedance* is more general than the word *resistance*, and is used where reactive components present. Normally expressions derived for resistive components are equally valid for impedances.

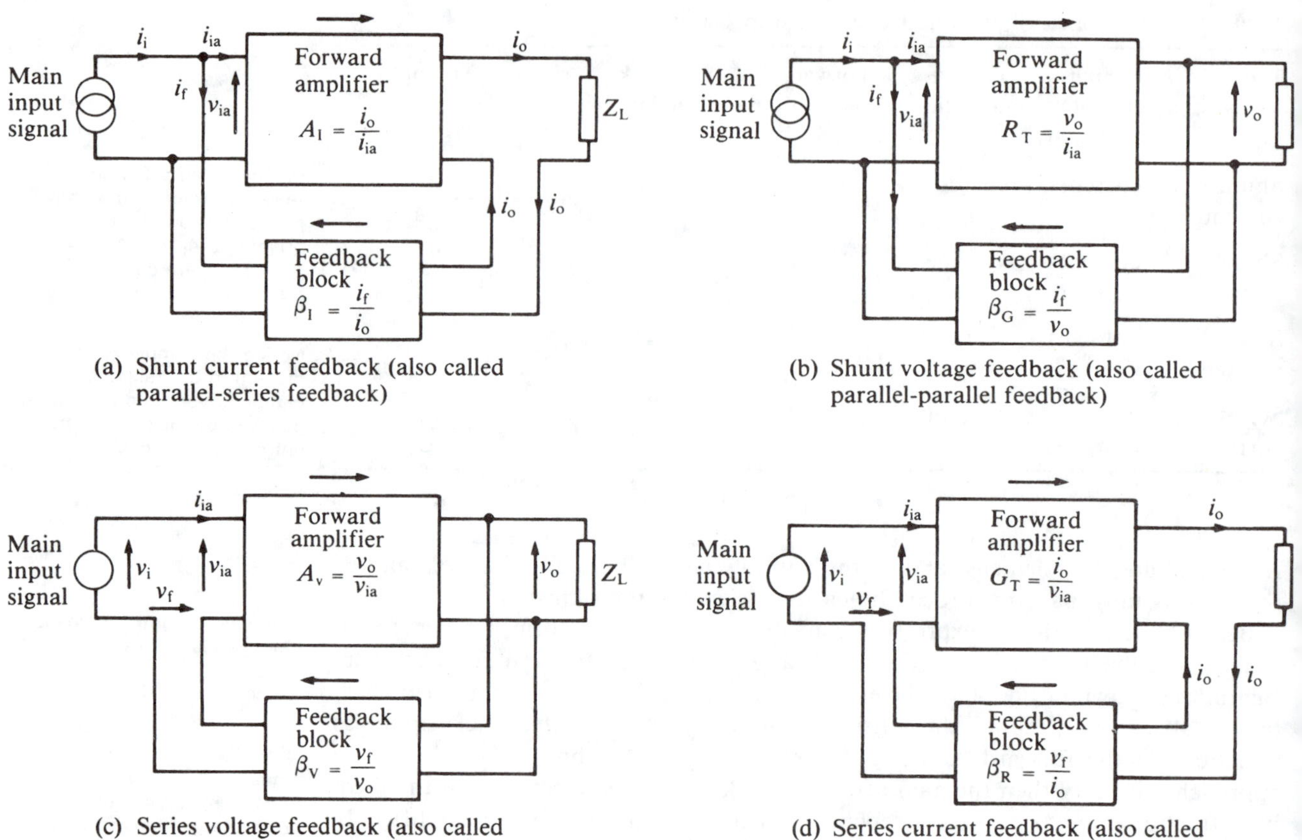

(a) Shunt current feedback (also called parallel-series feedback)

(b) Shunt voltage feedback (also called parallel-parallel feedback)

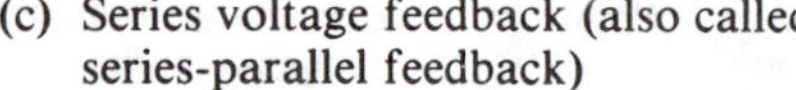

(c) Series voltage feedback (also called series-parallel feedback)

(d) Series current feedback (also called series-series feedback)

Fig. 4.1 The four feedback amplifier configurations.

This section shows that negative feedback applied in the four ways (Fig. 4.1) allows the input and output impedances of the forward amplifier to be increased or decreased as required.

First examine the effect on input impedances at the input side of the amplifier. Of the four configurations two have feedback applied in series at the input (series voltage and series current) and two have feedback applied in parallel at the input side (shunt current and shunt voltage). Figs. 4.2a and 4.2b show the input sides for series input and shunt input configurations respectively. Each diagram covers two separate cases of voltage and current feedback connection at the output. The effect of feedback on input impedance is the same in both cases; this is shown as follows.

Consider first the effect of feedback on input impedance for the series input configurations, Fig. 4.2a. The input voltage to the forward amplifier equals the difference of the main input signal and the feedback voltage. Thus

Kirchhoff's Voltage Law has been invoked here.

$$v_{ia} = v_i - v_f \tag{4.5}$$

This signal v_{ia} passes round the loop formed by the forward amplifier and the feedback block and in doing so is multiplied by $A\beta$. Therefore

$$v_f = A\beta . v_{ia} \tag{4.6}$$

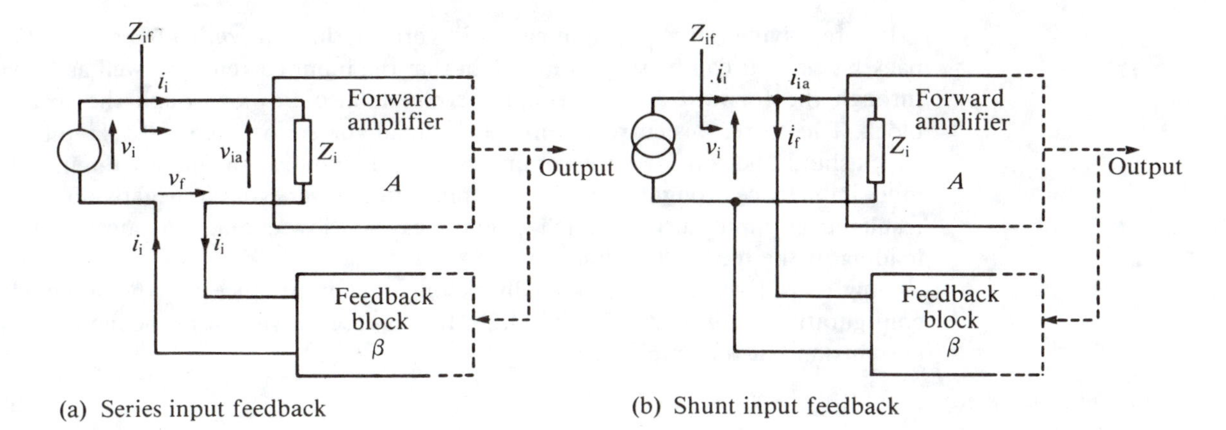

Fig. 4.2 Circuits for analysing the effect of feedback on input impedance.

For voltage output sampling the forward gain is A_V and the feedback fraction is β_V (see Table 4.1), so $A\beta = A_V.\beta_V$ in this case. For current output sampling $A\beta = A_G.\beta_R$.

Substituting Equation 4.6 for v_f in Equation 4.5 gives

$$v_{ia} = v_i - A\beta.v_{ia}$$

Rearranging this equation gives

$$v_i = v_{ia}(1 + A\beta)$$

The next step is to divide both sides by the input current i_i,

$$\frac{v_i}{i_i} = \frac{v_{ia}}{i_i}.(1 + A\beta) \qquad (4.7)$$

The reason for this step is clear when it is realised that the quotient v_i/i_i is equal to the input impedance Z_{if} as seen by the main signal source after feedback has been applied. The current i_i also flows through the input terminals of the forward amplifier and so the quotient v_{ia}/i_i is equal to the input impedance of the amplifier Z_i before feedback is applied. Thus Equation 4.7 becomes

$$Z_{if} = Z_i(1 + A\beta) \qquad (4.8)$$

Yet another example of the occurrence of the feedback factor $(1 + A\beta)$.

This shows that the effect of *series-input feedback* is to cause the input impedance of the forward amplifier to be *multiplied* by the feedback factor $(1 + A\beta)$, and for negative feedback has the effect of *increasing* the input impedance.

Worked Example 4.1

Series feedback is applied at the input of an amplifier which has an input resistance of 1 kΩ and a voltage gain of $A_V = 1000$. If the feedback fraction is $\beta_V = 0.1$ calculate the input resistance after feedback is applied.

Solution: We have $A\beta = A_V.\beta_V = 1000 \times 0.1 = 100$.

Therefore Equation 4.8 gives

$$Z_{if} = 10^3 \times (1 + 100) = 101 \text{ k}\Omega$$

Before leaving the series input case it is worth noting a *loading effect* on the feedback block β. It can be seen in Fig. 4.2a that the input current i_i as well as flowing through the forward amplifier input terminals also flows through the feedback block. Therefore, this current constitutes a load current on the feedback voltage V_f which should be considered when analysing real circuits. Fortunately the increased input impedance brought about by the application of series-input negative feedback, results in an input current which in many cases is low enough to cause negligible loading of the feedback voltage.

Now consider the effect of feedback on input impedance for the shunt input configurations represented by Fig. 4.2b. In this case the signals at the input side are currents and at the amplifier input

Kirchhoff's Current Law.

$$i_{ia} = i_i - i_f \tag{4.9}$$

Parallel connections and series connections are duals because in one currents are added, and in the other voltages are added. The steps in the following derivation therefore mirror those just completed for the series-input feedback case.

As in the series input case the amplifier input signal passes round the loop formed by the forward amplifier and the feedback block to give

$$i_f = A\beta i_{ia} \tag{4.10}$$

In the case of voltage output sampling then Table 4.1 shows that $A\beta = R_T.\beta_G$, while for current output sampling $A\beta = A_I\beta_I$. Substituting Equation 4.10 for i_f in Equation 4.9 gives

$$i_{ia} = i_i - A\beta i_{ia}$$

Rearranging,

$$\frac{1}{i_i} = \frac{1}{i_{ia}} \frac{1}{(1 + A\beta)}$$

and multiplying by v_i gives

$$\frac{v_i}{i_i} = \frac{v_i}{i_{ia}} \cdot \frac{1}{(1 + A\beta)} \tag{4.11}$$

The quotient v_i/i_i is again recognised as Z_{if}, the input impedance to the amplifier after feedback has been applied. Also, note that because the feedback is connected in parallel v_i also equals the input voltage to the amplifier and therefore the quotient v_i/i_{ia} is recognised as Z_i, the input impedance to the forward amplifier before feedback is applied. Therefore Equation 4.11 can be written in the form

$$Z_{if} = Z_i \frac{1}{1 + A\beta} \tag{4.12}$$

This shows that shunt-input feedback has the opposite effect on input impedance than does series-input feedback. With *shunt-input feedback* the amplifier input impedance is *divided* by the feedback factor $(1 + A\beta)$ which for negative feedback means that the input impedance is *reduced*.

Again there is a *loading effect* on the feedback circuit which should be noted. The input signal current causes an input voltage v_i which also appears across the output of the feedback block. This tends to drive a current through the feedback block which should be considered when analysing a real feedback circuit. Fortunately as with the series input arrangement, negative feedback helps because the lowered input impedance means that in practice the amplitude of v_i is low enough to permit the loading effect to be neglected in many cases.

Loading effects also occur at the output side of feedback amplifiers and the effects can be neglected in many practical cases. Nevertheless, loading effects cannot always be discounted. Methods for analysing feedback circuits where loading effects cannot be neglected are presented in Chapter 5.

Worked Example 4.2

Shunt feedback is applied at the input of an amplifier which has an input impedance of 10 kΩ and a trans-resistance of $R_T = 10^5\ \Omega$. If the feedback fraction is $\beta_G = 10^{-3}$ S, calculate the input impedance after feedback is applied.

Solution: We have $A\beta = R_T.\beta_G = 10^5 \times 10^{-3} = 100$. Therefore Equation 4.12 gives

$$Z_{if} = \frac{10\ \text{k}\Omega}{1 + 100} = 99.0\ \Omega$$

Having dealt with the effect of feedback on amplifier input impedance, now consider the effect of feedback on output impedance. The voltage-output and current-output configurations are represented in Figs 4.3a and b respectively.

Consider first the voltage-output configuration. The feedback circuit samples the output voltage and like a voltmeter ideally should draw no current from the circuit, $i_\beta = 0$. Therefore, the output voltage v_o and load current i_o are related by the equation

The approach is to manipulate the output equation after feedback to be same form as this, and thus allow Z_{of} to be identified.

$$v_o = e_o - Z_o i_o \tag{4.13}$$

where e_o is the no-load output e.m.f. of the amplifier and the term $Z_o i_o$ represents the voltage drop across the forward amplifier output impedance caused by load current flowing. If A is taken to be the gain of the amplifier with the load disconnected, then Equation 4.13 can be written as

For voltage-output amplifiers, *no-load* means that no output current flows and the load impedance is therefore assumed to be replaced by an open circuit.

$$v_o = A\,X_{ia} - Z_o i_o \tag{4.14}$$

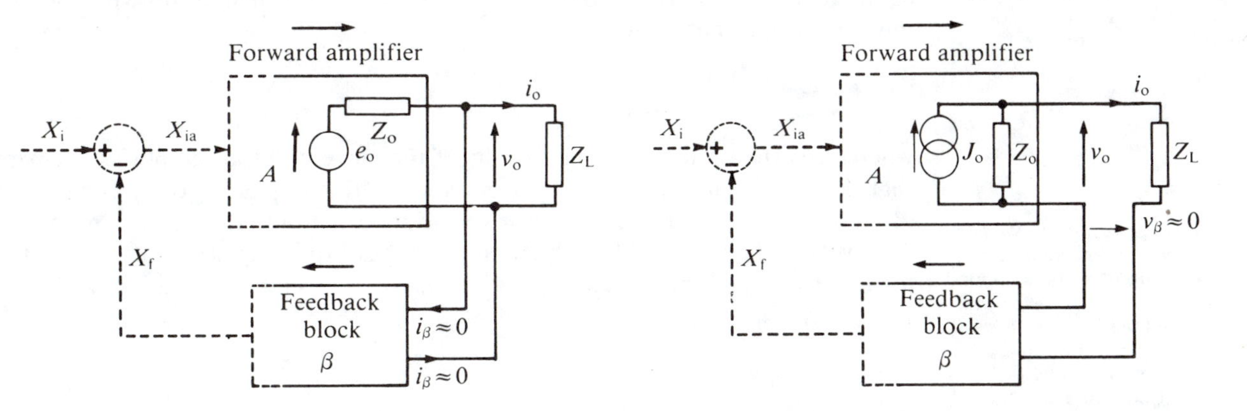

(a) Voltage output feedback (b) Current output feedback

Fig. 4.3 Circuits for analysing the effect of feedback on output impedance.

where for a current-input amplifier $X_{ia} = i_{ia}$ and $A = R_T$ while for a voltage-input amplifier $X_{ia} = v_{ia}$ and $A = A_V$.

Now X_{ia} is the input to the forward amplifier and not the main input signal to the feedback circuit. The latter is related to X_{ia} by the expression

$$X_{ia} = X_i - \beta.v_o \tag{4.15}$$

where for current input signals $\beta = \beta_G$ and for voltage input signals, $\beta = \beta_V$.

Substituting Equation 4.15 for X_{ia} in Equation 4.14 gives

$$v_o = A(X_i - \beta v_o) - Z_o i_o$$

and, rearranging,

Confirm this yourself.

$$v_o = \frac{A}{1 + A\beta} X_i - \frac{Z_o}{1 + A\beta} i_o \tag{4.16}$$

This equation for the closed-loop amplifier is now in the same form as Equation 4.14 for the forward amplifier output without feedback.

The second term on the right-hand side of this expression is recognised as the voltage drop across the output impedance of the amplifier after feedback has been applied. Therefore

$$Z_{of} = \frac{Z_o}{1 + A\beta} \tag{4.17}$$

which shows that for *voltage output feedback* the output impedance of the forward amplifier is *divided* by the feedback factor $(1 + A\beta)$, which for negative feedback results in a *lowering* of the output impedance. This is the same as the effect produced on the input impedance owing to feedback applied in shunt at the input side.

Because the output signal is a current the Norton equivalent circuit is used to represent the amplifier output.

Again because of duality the steps in the following analysis mirror those just carried out for the voltage-output case.

Lastly consider the current-output configuration shown in Fig. 4.3b. In this case the output current i_o is sensed by the feedback circuit which ideally should act as an ammeter and result in no voltage drop where it is inserted in the circuit, $v_\beta = 0$. Assuming this to be the case then the voltage across the amplifier output terminals is equal to the output voltage v_o across the load. In addition, the output current i_o is equal to the generator current J_o less the current which is diverted down the impedance Z_o. Therefore

$$i_o = J_o - \frac{v_o}{Z_o} \tag{4.18}$$

where J_o is the no-load output current of the current-output amplifier. Remember that *no load* for current generators means the opposite to that for voltage generators. For a current generator on no load the output is short-circuited. The term v_o/Z_o represents the load current down the output impedance Z_o. If A is the no load gain of the forward amplifier then $J_o = AX_{ia}$ and so Equation 4.17 may be written as

For a voltage output *no load* means zero load current, but for a current output as here no load means no voltage is developed across the load. Hence no load gain of a current output amplifier is obtained by taking the output current when the load impedance is replaced by a short circuit.

$$i_o = AX_{ia} - \frac{v_o}{Z_o} \tag{4.19}$$

where for a current-input amplifier $X_{ia} = i_{ia}$ and $A = A_I$, while for a voltage-input amplifier $X_{ia} = v_{ia}$ and $A = G_T$. Continuing the development along the same lines as for the previous voltage-output case then substituting Equation 4.15 for X_{ia} in Equation 4.19 gives

$$i_o = A(X_i - \beta i_o) - \frac{v_o}{Z_o}$$

and rearranging this provides the expression

$$i_o = \frac{A}{1 + A\beta} X_i - \frac{v_o}{Z_o(1 + A\beta)} \quad (4.20)$$

This shows that the effect of *current output feedback* is to *multiply* the forward amplifier output impedance by the feedback factor $(1 + A\beta)$. This means that for negative feedback the output impedance is *increased*.

In similar form to the equation for input impedance for current feedback, Equation 4.8.

In the above analysis of output impedance it has been assumed that the feedback block does not load the forward amplifier output circuit. Real circuits are never ideal and in practice i_β current (see Fig. 4.3) flows for the voltage output feedback and a non-zero voltage v_β is developed for current output feedback. As with the loading effects at the input side of the feedback circuit that were mentioned earlier, the output loading effects are often negligible but they should be borne in mind when analysing a feedback circuit.

The expressions for the effects of feedback on input and output impedances are shown in Table 4.2.

Table 4.2 Effect of Feedback on Input and Output Impedances

Type of feedback	Input impedance Z_{if}	Output impedance Z_{of}	$A\beta$
Shunt current	$\frac{Z_i}{1 + A\beta}$	$Z_o(1 + A\beta)$	$A_I.\beta_I$
Shunt voltage	$\frac{Z_i}{1 + A\beta}$	$\frac{Z_o}{1 + A\beta}$	$R_T.\beta_G$
Series voltage	$Z_i(1 + A\beta)$	$\frac{Z_o}{1 + A\beta}$	$A_V.\beta_V$
Series current	$Z_i(1 + A\beta)$	$Z_o(1 + A\beta)$	$G_T.\beta_R$

Note that in expressions for output impedance the gain constants A_I, R_T, A_V and G_T are obtained under no load conditions. These are open-circuit load for series-output feedback and short-circuit load for voltage output feedback.

It is not necessary to memorise this table because of the simple symmetry of its contents. If the connection is shunt-input/voltage-output then the input/output impedance is divided by the familiar factor $(1 + A\beta)$, while for series-input/current-output connection the input/output impedance is multiplied by $(1 + A\beta)$. The appropriate A and β parameters to use are easily determined once the input and output signals have been identified as being voltages or currents.

One way to remember that series feedback increases the impedance is that this also occurs when two resistances are connected in series. Conversely the reduction in impedance with parallel feedback also occurs when resistors are connected in parallel.

Exercise 4.2

Confirm the correctness of the last column of Table 4.2

It is now possible to justify the statement made earlier, that negative feedback tends to improve the type of amplifier to which it is applied. Consider what happens as the amount of feedback is increased. In the limit as $(1 + A\beta)$ reaches infinity the input and output impedances moves towards zero or infinity depending on the way feedback is applied, i.e. parallel or series. Also in the limit as $(1 + A\beta)$ reaches infinity the closed-loop gain becomes independent of the forward amplifier and equals $1/\beta$. Table 4.3 shows what happens to these quantities when the amount of negative feedback is infinitely large. Tables 4.1 and 4.2 have been used here. It is readily seen that in the limit each one of the feedback configurations behaves as an ideal amplifier or controlled source. For example, an ideal voltage amplifier, or voltage-controlled voltage source (v.c.v.s.) has a well-defined voltage transfer ratio and infinite input impedance and zero output impedance. This corresponds to the limiting case for series-voltage feedback. The last column in Table 4.3 lists the corresponding ideal amplifiers. This table confirms that negative feedback does indeed tend to improve the amplifier characteristics towards one of four ideal amplifiers.

Of course, in practice, the limit can never be reached, but by applying large amounts of negative feedback amplifiers can be made which are very nearly ideal.

Table 4.3 Limiting Case for Large Applied Negative Feedback

Type of feedback	Input impedance Z_{if}	Output impedance Z_{of}	Gain $(1 + \beta)$	Corresponding ideal amplifier
Shunt current	zero	infinite	$\frac{i_o}{i_i} = \frac{1}{\beta_I}$	current amplifier (c.c.c.s.)
Shunt voltage	zero	zero	$\frac{v_o}{i_i} = \frac{1}{\beta_G}$	trans-resistance amplifier (c.c.v.s.)
Series voltage	infinite	zero	$\frac{v_o}{v_i} = \frac{1}{\beta_V}$	voltage amplifier (v.c.v.s.)
Series current	infinite	infinite	$\frac{i_o}{v_i} = \frac{1}{\beta_R}$	trans-admittance amplifier (v.c.c.s.)

Shunt-Voltage Feedback Circuit Example

In this and the remaining sections of this chapter circuit implementations of the four feedback configurations are discussed. These should be looked on as typical examples which illustrate how negative feedback principles can be implemented in circuit form; many other circuits could have been chosen.

It is good practice for a circuit designer to collect circuit ideas and expand his repertoire. Collections of electronic circuits are published in book form and can be very useful. See for example Markus, J. *Modern Electronics Circuits Reference Manual* (McGraw-Hill, 1980) and other books by the same author.

Commencing with the shunt-voltage negative feedback configuration because this turns out to be one of the simplest: To form a feedback circuit a suitable forward amplifier and feedback block have to be chosen. Reference to Fig. 4.1b shows that the shunt-voltage configuration operates in trans-resistance mode with the input signal being a current i_i, and the output signal a voltage. Therefore, a forward amplifier is required which operates in the same trans-resistance mode. The feedback fraction for this configuration is $\beta_G = i_f/v_o$ and therefore the feedback block is required to operate in trans-conductance mode.

One of the properties of negative feedback is that it reduces the sensitivity to variations in forward amplifier gain and closed-loop gain is mainly defined by the feedback. Because of manufacturing spreads in active device parameters it is generally desirable to form the feedback block from passive components. The feed-

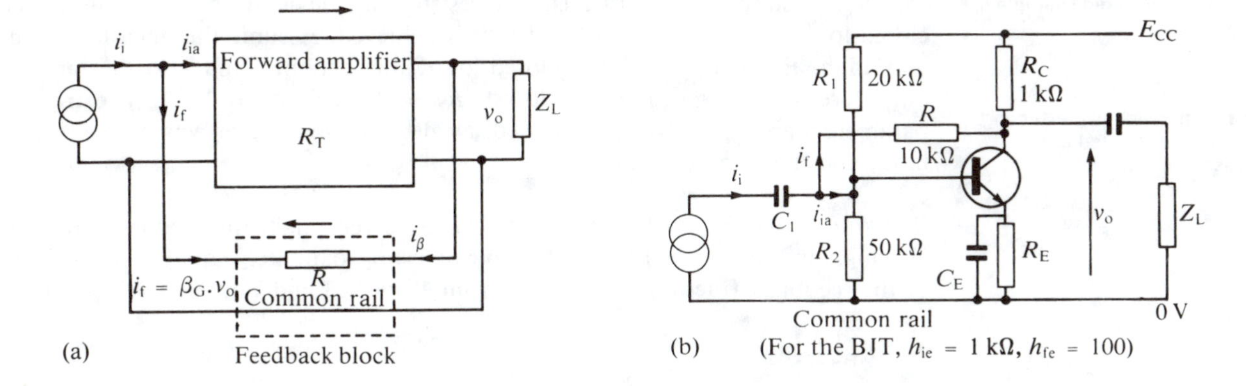

Fig. 4.4 Shunt voltage feedback example: (a) block diagram, (b) BJT circuit.

back block for the present shunt-voltage configuration is required to sense the output voltage v_o, and convert this to a current i_f. A very simple but effective way to do this is by means of a single resistor R, as shown in Fig. 4.4a. In normal operation the voltage at the point where R joins the input to forward amplifier is quite low because the effect of shunt-input negative feedback is to cause the input impedance to be small at that point. Therefore most of the output voltage v_o at the other end of R is developed across R. To a good approximation the current through the resistor is given by Ohm's law,

$$i_\beta \approx \frac{v_o}{R} \tag{4.22}$$

The diagram shows that the direction of flow of i_β is opposite to that marked for i_f and so

$$\beta_G = \frac{i_f}{v_o} = \frac{-i_\beta}{v_o} = -\frac{1}{R} \tag{4.23}$$

For the forward amplifier a negative trans-resistance parameter is required. This is because β_G is negative and to achieve a negative feedback system the loop gain $(1 + R_T.\beta_G)$ must be greater than unity. Therefore, the forward amplifier must be phase inverting. A helpful feature of this configuration is that a common line runs from the lower input terminal of the forward amplifier, through the feedback circuit and to the lower terminal of the amplifier output. A simple forward amplifier which is single-ended at input and output can therefore be used, such as a single-stage common-emitter BJT amplifier. The full feedback circuit is shown in Fig. 4.4b. In this circuit the feedback is via R and other components are the normal biasing resistors and coupling and decoupling capacitors. The presence of R causes a small quiescent current to flow from collector to base which can be taken account of when calculating the circuit biasing. Alternatively the current can be blocked by inserting a capacitor in series with the feedback resistor R provided it is large enough to have a reactance which is negligibly small compared with R at the signal frequencies of interest.

The design of the biasing arrangements and choice of components is covered in texts on transistor circuits. See, for example, Ritchie, G.J. *Transistor Circuit Techniques*, Second edition, Van Nostrand Reinhold (International), 1987.

Worked Example 4.3 Calculate a suitable value of feedback resistor for the circuit of Fig. 4.4b so that a closed-loop trans-resistance of $-10\ \text{k}\Omega$ is achieved. Assume the open-loop gain is very high. Estimate the actual closed-loop gain obtained and also the input and output resistance, given $R_1 = 20\ \text{k}\Omega$, $R_2 = 50\ \text{k}\Omega$, $R_c = 1\ \text{k}\Omega$. The transistor parameters are $h_{ie} = 1\ \text{k}\Omega$, $h_{fe} = 100$. Neglect any loading effects of the feedback block and the external load Z_L.

Cases where loading effects cannot be neglected are discussed in Chapter 5.

Solution: If the forward gain is assumed to be very high, the feedback fraction is also high and therefore the closed-loop gain is approximately equal to the reciprocal of the feedback fraction. Therefore, from Table 4.3 and Equation 4.23

$$R_{Tf} \approx \frac{1}{\beta_G} = -R$$

and so $R = 10\ \text{k}\Omega$.

To see what actual performance can be expected, first calculate the parameters of the forward amplifier on open-loop, with R removed. From basic transistor circuit theory,

For any three resistors R_A, R_B, R_C in parallel, first combine any two (say $R_A \| R_B$) and then combine the result with the third: $(R_A \| R_B) \| R_C$. Alternatively use the form $(R_A^{-1} + R_B^{-1} + R_C^{-1})^{-1}$. Both give the same result of course.

$$\text{Input resistance } r_i = R_1 \| R_2 \| h_{ie} = 20\ \text{k}\Omega \| 5\ \text{k}\Omega \| 1\ \text{k}\Omega \| = 800\ \Omega$$

$$\text{Output resistance } r_o = R_c = 1\ \text{k}\Omega$$

Further,

$$i_b = i_i \times \text{Input current coupling factor}$$

$$= i_i \times \frac{R_1 \| R_2}{(R_1 \| R_2) + h_{ie}} = i_i \cdot \frac{4\ \text{k}\Omega}{4\ \text{k}\Omega + 1\ \text{k}\Omega} = 0.8\ i_i$$

and $v_o = -R_c i_c = -R_c h_{fe} i_b$. Therefore $v_o = -R_c h_{fe}\,(0.8\ i_i)$ and so

$$R_T = \frac{v_o}{i_i} = -0.8\ R_c h_{fe} = -0.8 \times 1000 \times 100 = -80\ \text{k}\Omega$$

The feedback factor can now be calculated:

$$\text{Feedback factor} = 1 + R_T.\beta_G = 1 + 8 \times 10^4 \times \frac{1}{10^4} = 9$$

From which (see Tables 4.1 and 4.2)

$$\text{Closed-loop gain, } R_{Tf} = \frac{R_T}{1 + R_T\beta_G} = \frac{-80\ \text{k}\Omega}{9} = -8.89\ \text{k}\Omega$$

$$\text{Closed-loop input resistance } r_{if} = \frac{r_i}{1 + R_T\beta_G} = \frac{800\ \Omega}{9} = 89\ \Omega$$

$$\text{Closed-loop output resistance } r_{of} = \frac{r_o}{1 + R_T.\beta_G} = \frac{1000\ \Omega}{9} = 111\ \Omega$$

The value of R is obtained by solving the closed-loop expression $R_{Tf} = R_T/(1 + R_T\beta_G)$ for β_G and then using Equation 4.23 to obtain R.

As expected in the above example negative feedback has reduced both input and output resistances. The resulting gain is a little lower than the desired value of $-10\ \text{k}\Omega$ because the amount of negative feedback as indicated by a feedback factor of 9, is not very large. This could be compensated for by raising the value of R.

Show that for the circuit in the previous example if $R = 11.4\ \text{k}\Omega$ the desired closed-loop trans-resistance of $-10\ \text{k}\Omega$ is obtained.

Exercise 4.3

In Worked Example 4.3 the loading effect due to feedback has been neglected and so the calculated values of the closed-loop parameters, R_{Tf}, r_{if}, r_{of} are different from the true values. At the input the resistor R causes some of the forward signal to be diverted away from the base of the transistor in a similar way to the biasing resistors R_1 and R_2, thus lowering the forward gain. However, the value of feedback resistor R is ten times the value of transistor input resistance h_{ie} and so the effect is not large. Similarly, at the amplifier output R diverts some of the signal current at the collector away from the collector resistor R_c. Again the lowering effect on the gain is not large because the feedback resistor is ten times R_c in value. Where this loading effect is significant, or where it is important to have greater accuracy in calculations, then the methods described in the next chapter can be used.

Note how these two ratios of ten arise from the values of 100 given for h_{fe} — they would be increased for higher-gain transistors.

Series-Voltage Feedback Circuit Example

For the series-voltage feedback arrangement (Fig. 4.1c) both input and output signals are voltages. The forward amplifier works in the voltage amplification mode. The feedback block is required to sample the output voltage and return a fraction of it to the input. For the feedback block a simple passive circuit is desired to produce a voltage equal to a constant fraction of the input to it (the input to the feedback block being the output of the complete feedback circuit). A very simple circuit to do this is the potential divider circuit using two resistors connected in series. The series-voltage configuration with a potential divider for the feedback block is shown in Fig. 4.5a. The potential divider is a very common circuit. It can be seen in the treatment of voltage coupling factors in Chapter 3. Using the results obtained there it can be seen that R_1 in Fig. 4.5a is the equivalent source resistance to the signal v_o and R_2 is the equivalent load resistance. Also the feedback fraction β_V is equivalent to the voltage coupling factor and so by Equation 3.14,

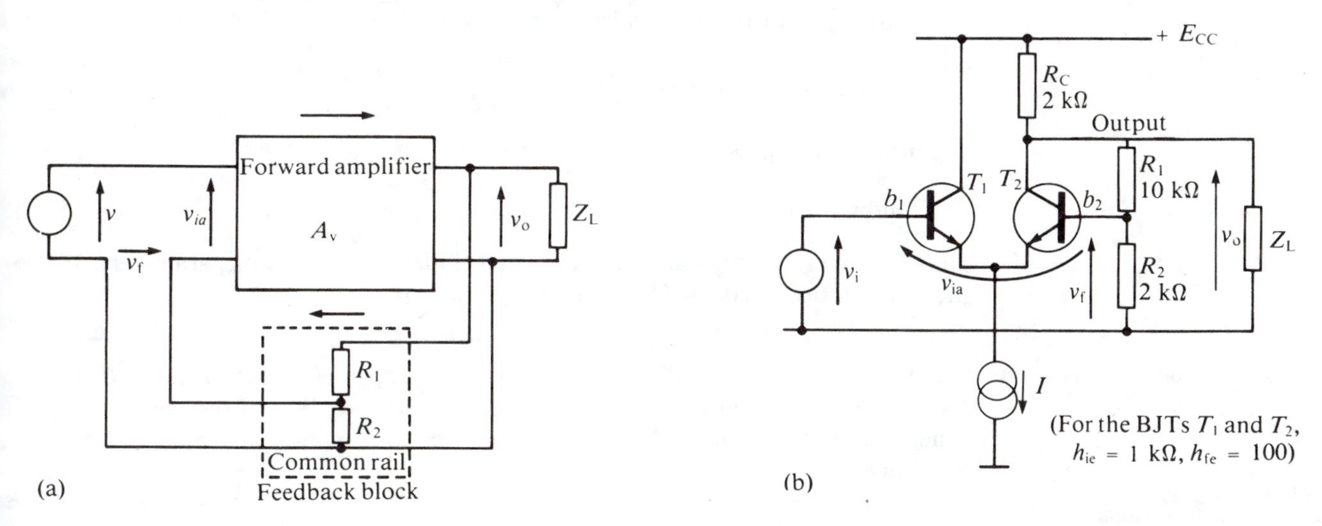

Fig. 4.5 Series-voltage feedback example: (a) block diagram, (b) BJT circuit.

$$\beta_V = \frac{R_2}{R_1 + R_2} \tag{4.24}$$

Exercise 4.4 Show that if the amount of negative feedback is large the closed-loop gain of Fig. 4.5a is given approximately by

$$A_{V_f} \approx 1 + \frac{R_1}{R_2}$$

Thus in this configuration the feedback fraction β_V is positive. Therefore a forward amplifier must be chosen with positive voltage gain. The common rail for the circuit runs from the lower terminal of the signal source, through the feedback network and joins the lower terminal of the forward amplifier output. Therefore, a single-ended output amplifier can be used. However, neither of the two amplifier input terminals joins directly to the common rail. One terminal connects to the main signal input and the other is connected to the output of the feedback block. Therefore an amplifier with single-ended input cannot be used as in the case of the shunt-voltage configuration, and an amplifier with differential inputs is required. A well-known BJT differential amplifier is the two-transistor *long-tailed pair*. This is shown in Fig. 4.5b and comprises transistors T_1, T_2, resistance R_c, and constant current source I. (In practice the constant current tail, I, is often provided by the collector of a third transistor or by a high-value resistor connected to a negative voltage.) A detailed explanation of the operation of the long-tailed pair circuit is to be found in texts on transistor circuits.

See, for example, Ritchie, G.J. *Transistor Circuit Techniques*, Second edition, Van Nostrand Reinhold (International), 1987.

The circuit amplifies the difference in the voltages applied to the bases, b_1 and b_2, of the two transistors. This difference voltage is marked as v_{ia} on Fig. 4.5b and corresponds to the terminal of the forward amplifier in Fig. 4.5a (corresponding to b_1 of transistor T_1) which comes directly from the main input voltage. The voltage at the lower terminal of the forward amplifier (corresponding to b_2 of transistor T_2) comes from the potential divider formed by R_1 and R_2, which in turn is connected across the output of the forward amplifier (this corresponds to the collector of transistor T_2). For the long-tailed pair the following relationships apply.

$$\text{Differential voltage gain, } A_{dm} = +\frac{1}{2} h_{fe} \cdot \frac{R_c}{h_{ie}} \tag{4.25}$$

$$\text{Input resistance, } r_i = 2\,h_{ie} \tag{4.26}$$

$$\text{Output resistance, } r_o = R_c \tag{4.27}$$

Here the two transistors are assumed to have identical transistor parameters. In the present notation used for feedback configurations A_{dm} is equal to A_V.

Worked Example 4.4 For the series-voltage feedback circuit in Fig. 4.5b, $h_{ie} = 1\ \text{k}\Omega$, $h_{fe} = 100$, $R_c = 2\ \text{k}\Omega$, $R_1 = 10\ \text{k}\Omega$ and $R_2 = 2\ \text{k}\Omega$. Neglecting loading effects of the feedback components and Z_L calculate the voltage amplification, input resistance, and output resistance.

It is a wise practice to make quick estimates where possible because they provide a check on calculations.

Solution: A quick estimate of what to expect for the amplification can be obtained by assuming the loop gain to be large and from Table 4.3

$$A_{Vf} \approx \frac{1}{\beta_V} = \frac{R_1 + R_2}{R_1} = 1 + \frac{R_2}{R_1} = 1 + \frac{10{,}000}{2{,}000} = 6$$

The forward voltage gain of the amplifier is given by Equation 4.25

$$A_V = \frac{1}{2} \times 100 \times \frac{2\ \text{k}\Omega}{1\ \text{k}\Omega} = +100$$

Thus using Table 4.2 the feedback factor is given by

$$(1 + A_V \cdot \beta_V) = (1 + 100 \times \frac{R_1}{R_1 + R_2}) = (1 + 100 \times \frac{2\ \text{k}\Omega}{2\ \text{k}\Omega + 10\ \text{k}\Omega})$$

$$= 17.67$$

Therefore the closed-loop gain is

$$A_{V_f} = \frac{A_V}{1 + A_V \cdot \beta_V} = \frac{100}{17.67} = 5.66$$

This accords with the initial estimate of $A_{Vf} \approx 6$.

Because the feedback is series input, the input resistance is multiplied by the feedback factor and so

$$r_{if} = r_i(1 + A_V \cdot \beta_V) = (2 \times 1\ \text{k}\Omega)\,(17.67) = 35.3\ \text{k}\Omega$$

The equation for r_{if} and the following one for r_{of} are obtained from Table 4.2.

Finally the output resistance with voltage-output feed is divided by the feedback factor and the closed-loop value is calculated as follows:

$$r_{of} = \frac{r_o}{(1 + A_V \cdot \beta_V)} = \frac{2\ \text{k}\Omega}{17.67} = 113\ \Omega$$

The series-voltage feedback has provided a circuit with increased input resistance and reduced output resistance.

Exercise 4.5

Assuming that loading effects can be neglected, show that the closed-loop voltage gain of the circuit in Fig. 4.5a is determined by the relative values, and not the absolute values of R_1 and R_2.

Series-Current Feedback Circuit Example

The input signal to the forward amplifier in the series-current feedback configuration is a voltage. The output signal is a current. The feedback block has to sample the output current and feed a fraction of this to the input as a voltage. A simple way to do this is to insert a resistor R in the output loop. The output current in passing through R develops a voltage across it which is proportional to the output current, and this voltage is then passed to the input side of the amplifier where it is subtracted from the main input signal. Fig. 4.6a shows the feedback arrangement with this type of feedback block. If the current gain of the amplifier is high, the current circulating at the input side of the amplifier is small and presents negligible loading of the feedback block. Then v_f is given by Ohm's law; $v_f = -R \cdot i_o$. The negative sign occurs because the direction of flow of i_f produces an ohmic resistor voltage $R \cdot i_f$ which is opposite in polarity to that for the convention adopted for v_f in the diagram. Therefore,

Taking up the point of duality again: It is interesting to see that in this series-current case the feedback network consists of a resistor R connected across (i.e. in parallel) the conductors of the feedback block. This is the exact opposite of the shunt-voltage case (Fig. 4.5a) where R is connected in series in the feedback network.

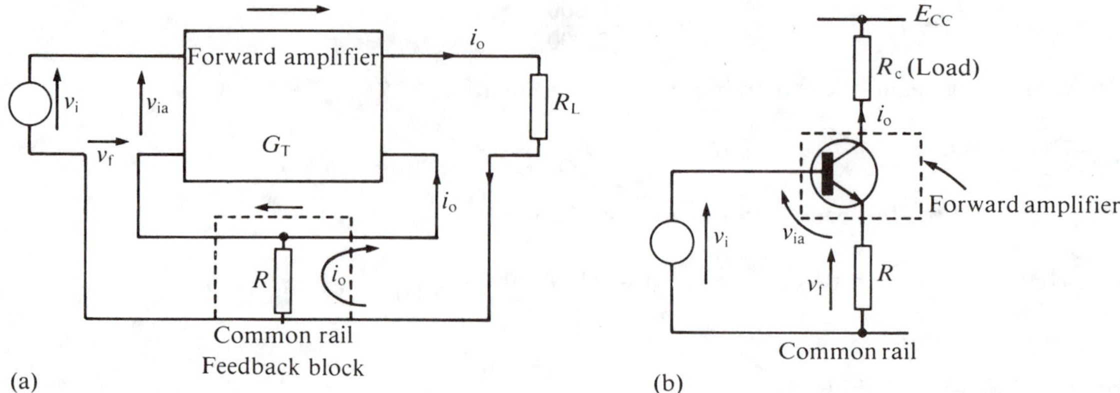

Fig. 4.6 Series-current feedback example: (a) block diagram, (b) BJT circuit.

$$\beta_R = \frac{v_f}{i_o} = -R \tag{4.28}$$

The overall feedback circuit has a single-ended input and output. This can be seen from Fig. 4.6a where there is a direct common rail connection running from the lower terminal of the main signal source, through the feedback block and finishing on the lower terminal of the load. It can also be seen that none of the terminals of the forward amplifier are connected to the common rail. Therefore, the forward amplifier has to be a *floating* circuit. It turns out that this somewhat restricts the range of active circuits from which to choose a forward amplifier when compared with the two previously considered configurations, shunt-voltage and series-voltage, where at least one side of the forward amplifier is single ended. However, examples are not hard to come by. Because β_R is negative, see Equation 4.28, a negative forward amplifier gain G_T is required if the feedback is to be negative. Conveniently a single common-emitter BJT can act as a floating forward amplifier with negative gain as shown in the series-current feedback circuit in Fig. 4.6b. For simplicity the usual bias circuitry is not shown. This circuit is a well known basic transistor circuit in which the resistor in the emitter connection is used to provide increased input impedance and to stabilize the collector current. Transistor circuit analysis techniques provide the result

All active devices, thermionic-valves, FETs, BJTs, have this property and the resistance R gives *negative* feedback.

$$r_{if} = h_{ie} + (1 + h_{fe})R \tag{4.29}$$

$$G_{Tf} = \frac{i_o}{v_i} = \frac{-h_{fe}}{h_{ie} + (1 + h_{fe})R}\ \text{S} \tag{4.30}$$

$$r_{of} = \infty\ \Omega \tag{4.31}$$

where r_{of} is the output resistance seen by the load r_L, and the only transistor parameters of importance are assumed to be h_{ie} and h_{fe}. In the following worked example the circuit is analysed using feedback theory instead.

Worked Example 4.5 Using feedback theory derive expressions for input resistance, trans-conductance, and output resistance of the circuit shown in Fig. 4.6b.

Solution: The forward amplifier consists of the transistor operating in common-emitter mode. Therefore, assuming $h_{re} = h_{oe} = 0$, the output resistance of the forward amplifier is infinity and the input resistance is equal to h_{ie}. The current gain is given by

$$h_{fe} = \cdot\frac{i_c}{i_b} = \frac{-i_o}{i_b}$$

Therefore,

$$G_T = \frac{i_o}{v_i} = \frac{i_o}{(i_b\, h_{ie})}$$

Hence

$$G_T = \frac{1}{h_{ie}}\cdot\frac{i_o}{i_b} = \frac{-h_{fe}}{h_{ie}}$$

These expressions for G_T, r_i and r_o are for the parameters before feedback is applied. From these the closed-loop values can be obtained.

Also $r_i = h_{ie}$ and $r_o = \infty$.

The feedback fraction is given by Equation 4.28, $\beta_R = -R$. Therefore, after feedback is applied, Tables 4.1 and 4.2 give the following relationships. For input resistance:

$$r_{if} = r_i(1 + G_T\cdot\beta_R) = h_{ie}\left(1 + \frac{-h_{fe}}{h_{ie}}(-R)\right)$$

Therefore

$$r_{if} = h_{ie} + h_{fe}R \quad \Omega \qquad (4.32)$$

The gain after feedback is given by

$$G_{Tf} = \frac{G_T}{1 + G_T\beta_R} = \frac{\dfrac{-h_{fe}}{h_{ie}}}{1 + \dfrac{-h_{fe}}{h_{ie}}(-R)}$$

Therefore

$$G_{Tf} = \frac{-h_{fe}}{h_{ie} + h_{fe}R}\ \mathrm{S} \qquad (4.33)$$

Finally

$$r_{of} = r_o(1 + G_T\beta_R) = \infty\left(1 + \frac{-h_{fe}}{h_{ie}}(-R)\right)$$

Therefore

$$r_{of} = \infty \quad \Omega \qquad (4.34)$$

Discrepancy.

A comparison of Equations 4.29 and 4.32 for input resistance after feedback and also Equations 4.30 and 4.33 for trans-resistance after feedback shows a discrepancy. The factor $(1 + h_{fe})$ in the exact Equation 4.29 and 4.30 appears as h_{fe} in the corresponding Equations 4.32 and 4.33 obtained using feedback theory. The reason for the discrepancy lies in the assumption in the feedback theory that the

And the reason for the discrepancy.

feedback block is not loaded by the small current which circulates in the input side of the feedback circuit and passes through the feedback block. In fact, the input current passes through the transistor emitter and so the current passing through resistor R is the true output current through the collector plus a small current from the input. Typically $h_{fe} = 100$ and the discrepancy is usually not important, especially when the manufacturing variability of h_{fe} is considered. However, where it is desired to take loading effects into account then the methods described in the next chapter can be used.

Shunt-Current Feedback Circuit Example

Once again raising the point about duality we see that the feedback block is the reverse of that for the opposite series-voltage feedback circuit, Fig. 4.5a.

This feedback configuration is the one shown in Fig. 4.1a. The feedback block is inserted in series with the output loop and therefore samples the output current. The feedback block output is connected in shunt at the input side of the amplifier and so the signal feedback is also a current. A simple circuit to sample the output current and feed back a fraction of it back as a current can be made from two resistors, and is shown incorporated in the feedback block shown in Fig. 4.7a. The two resistors are labelled R_1 and R_2. To understand how this feedback block functions, start as before by assuming that the loading effect on the feedback block by the forward amplifier input is negligible. In this case because the feedback signal is a current this means it is assumed that the low-input impedance owing to the shunt feedback at the input side of the amplifier results in a negligibly small voltage being developed across the output terminals of the feedback circuit. Hence, the left-hand side of R_1 is approximately at common rail potential. The lower terminal of R_2 is also at common rail potential. The other terminals of R_1 and R_2 are joined together. This means that both R_1 and R_2 develop approximately the same voltage across each other. They are effectively in parallel as viewed by the output current i_o which flows round the output loop of the feedback configuration. The voltage v_β is therefore given by

$$v_\beta = (R_1 \| R_2) i_o = \frac{R_1 R_2}{R_1 + R_2} \cdot i_o \tag{4.35}$$

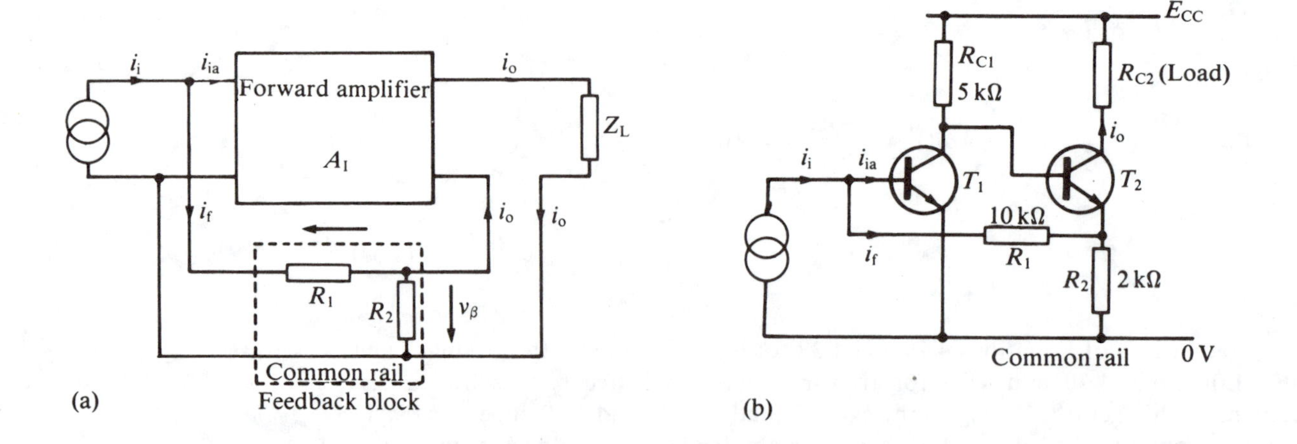

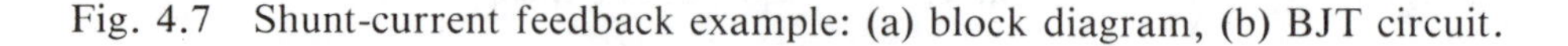
Fig. 4.7 Shunt-current feedback example: (a) block diagram, (b) BJT circuit.

Now because the left hand terminal of R_1 is approximately at common rail potential the current through R_1 is v_β/R_1. This current is equal to the feedback current i_f, and using Equation 4.35

$$i_f = \frac{v_\beta}{R_1} = \frac{1}{R_1} \cdot \frac{R_1 R_2}{R_1 + R_2} i_o = \frac{R_2}{R_1 + R_2} i_o$$

By analogy with the potential divider this circuit is often called a current divider. Note that this arrangement of resistors can also be viewed as providing current coupling from i_o to i_f, and the expression for current coupling factor Equation 3.16 in the previous chapter, can be used to obtain the same equation for β_I as given here.

Therefore the feedback fraction is given by

$$\beta_I = \frac{i_f}{i_o} = \frac{R_2}{R_1 + R_2} \qquad (4.36)$$

Equation 4.36 shows that β_I is positive and to obtain a negative feedback arrangement the forward amplifier has to provide a current gain which is also positive. One of the input terminals to the forward amplifier can be seen from Fig. 4.7a to be at common rail potential. However, neither terminal at the output of the forward amplifier is at common rail potential and so whatever circuit is chosen must have a floating output. A forward amplifier which meets this requirement is shown in Fig. 4.7b together with feedback components R_1 and R_2. This circuit is the one commonly used to implement shunt-current feedback. The forward amplifier comprises two phase-inverting common-emitter stages to provide an overall current gain which is positive, as required. The floating output for the forward amplifier is obtained by inserting one of the feedback resistors R_2 into the emitter lead of a second transistor. This idea has also been used in the series-current feedback example shown in Fig. 4.6b which also requires a floating output for the forward amplifier. The present circuit is also like that of Fig. 4.6b in that the feedback circuit does not sample the actual output signal current. The feedback circuit samples the emitter current of the second transistor whereas the true output current is the collector current of the same transistor (or, to be more exact, the negative of the collector current because the conventional direction for i_o is opposite to that for collector current). However, the error introduced by this is small because the emitter and collector currents in practice are close in value. From basic transistor theory, $i_c = i_e h_{fe}/(1 + h_{fe})$, and therefore for a modern transistor with $h_{fe} = 100$ or more the collector and emitter currents are equal to within 1% or less.

Exercise 4.6

Show that if the amount of negative feedback is large in Fig. 4.7a, the closed-loop gain is given approximately by

$$A_{If} \approx 1 + \frac{R_1}{R_2}$$

The circuit differs from the other feedback circuit examples examined in the loading effects owing to the feedback block which, in practice, cannot be neglected. It can be seen that in Fig. 4.7b feedback resistor R_2 is in the emitter lead of the second transistor. This resistor as well as sensing the output current to provide overall feedback via R_1, also provides local feedback to the second transistor. The resistor R_2 provides local series-current feedback to the second transistor in the same manner as shown in Fig. 4.6b. This complicates the analysis of the circuit and for this reason the analysis is deferred until the next chapter where loading effects in feedback circuits due to the feedback block are considered in more detail.

General points about loading in feedback circuits.

Other loading effects not considered so far are those owing to the load impedance connected at the output of the feedback circuit, and also to finite self impedances of the main signal source. In general, the effect of the load impedance is to reduce the coupling factor at the output of the forward amplifier thus lowering the forward gain and reducing somewhat the beneficial effects of the negative feedback. The self-impedance of the main signal source can have the effect of reducing the effective amplitude of the feedback signal, v_f of i_f, that is returned to the input side of the forward amplifier. This also can reduce the beneficial effects of the negative feedback. These loading effects are also dealt with in the next chapter.

Summary

In a feedback amplifier, because signals can be considered as either currents or voltages the feedback block can sense the output signal in two ways and apply a fraction of it to the amplifier input also in two ways. This leads to four feedback configurations: *shunt-current, shunt-voltage, series-voltage, series-current*. The basic feedback expression

$$A_f = \frac{A}{1 + A\beta}$$

applies to each of these configurations when the appropriate input-output ratio is used for A and β (see Table 4.1).

The effect of feedback is to change the input and output impedances of the amplifier circuit. If the feedback is series input, the input impedance is multiplied by the feedback factor $(1 + A\beta)$ and is increased when the feedback is negative. For shunt-input the input impedance is divided by $(1 + A\beta)$ and is therefore reduced. The output impedance is similarly increased by the factor $(1 + A\beta)$ for current-output feedback and reduced for voltage-output feedback. Table 4.2 gives the relationships. These properties are very useful because they permit amplifier input and output impedances to be improved in any direction to provide better matching of source impedances and load impedances to the amplifier.

For large amounts of negative feedback the closed-loop behaviour of the amplifier approaches that of an ideal amplifier. Each feedback configuration approaches one of the four ideal controlled sources. Which one it is can be determined by choosing how the feedback is applied. For example, voltage-output feedback senses the output voltage and it is the signal voltage behaviour that is improved. That is to say the output of the amplifier behaves as a controlled voltage generator with low self-impedance. On the other hand in shunt-input feedback a signal current is fed back and it is the current input behaviour which is improved. In other words, the amplifier is controlled by the input current and has a low input impedance.

Some typical negative feedback circuits have been presented which illustrate each of the four feedback configurations. Depending on the circuitry a feedback block can have a feedback fraction β which is either positive or negative. To obtain negative feedback the forward amplifier gain should have the same sign as the feedback fraction. Thus some configurations have forward amplifiers which have an even number of inverting transistor stages and others have an odd number of transistor stages. Of course amplifying circuits using field-effect transistors, or even integrated circuit amplifiers, can be used for the forward amplifier.

Current-output feedback circuits are not quite as convenient to implement as voltage-output feedback circuits because of the requirement to have a forward amplifier with floating output. Many modern applications of feedback principles use integrated circuit *operational amplifiers*, which usually do not have floating outputs and consequently they are commonly operated with voltage-output feedback. Operational amplifier applications are dealt with as a separate subject in later chapters.

From the discussion of feedback circuit examples has emerged the possibility of *loading effects* between the forward amplifier and feedback block. Loading effects can also be due to the signal source impedance and the load impedance connected at the feedback circuit output. In this chapter these effects have been assumed to be negligible. This assumption is valid for most practical cases. However, there can be occasions when it is not. It is often wise to analyse a feedback circuit with loading effects taken into account to verify whether they can be neglected. The analysis of feedback circuits with loading effects is dealt with in the next chapter.

Problems

In the following problems assume that loading effects are negligible.

4.1 Shunt-voltage feedback is applied to a forward amplifier using a feedback resistor of 10 kΩ. The forward amplifier has a trans-resistance parameter $K_R = -500$ kΩ, and input and output resistances both equal to 500 Ω. Calculate the gain, and input and output impedances after feedback is applied.

4.2 An amplifier has a voltage amplification of 1000 and input and output resistances of 10 kΩ and 1 kΩ respectively. Series-voltage feedback is applied as shown in Fig. 4.5a. Resistor values are $R_1 = 10$ kΩ, and $R_2 = 500$ Ω. Calculate the closed-loop input and output resistances, and also the gain.

4.3 For the series-current feedback circuit in Fig. 4.6b the transistor parameters are $h_{ie} = 1500$ Ω and $h_{fe} = 300$. What value of feedback resistor R, is required to give a closed-loop trans-conductance of -10^{-3} S? Calculate the input resistance of the circuit with this value of R.

4.4 A current amplifier with gain constant $K_I = 800$ and $r_i = 1$ kΩ and $r_o = 5$ kΩ is used in the feedback arrangement shown in Fig. 4.7a. The feedback resistors are $R_1 = 10$ kΩ and $R_2 = 800$ Ω. Calculate the closed-loop gain and input and output resistances after feedback is applied.

4.5 An amplifier having parameters $K_V = 1000$, $r_i = 10$ kΩ, $r_o = 1$ kΩ is used in the series-voltage feedback arrangement shown in Fig. 4.5a. Feedback is applied so that the sensitivity of the circuit amplification to variations in forward gain is reduced by a factor of 20. If $R_2 = 1$ kΩ calculate R_1. Also calculate the closed-loop gain and input and output resistances after feedback.

4.6 Two identical amplifiers having parameters $K_I = -50$, $r_i = 1$ kΩ, $r_o = 10$ kΩ, are connected in cascade. Overall shunt-current feedback is applied using the feedback arrangement shown in Fig. 4.7a with $R_1 = 1$ kΩ, $R_2 = 100$ Ω. Calculate A_{If}, r_{if} and r_{of}.

4.7 Shunt-voltage feedback is applied to an amplifier having a forward trans-resistance of $R_T = 10^5$ Ω. Feedback is applied by means of a resistor of value 2 kΩ. The input impedance of the amplifier is complex, and before feedback is applied it has a value of $100 - j500$ Ω. Calculate the real and imaginary parts of the input impedance after feedback is applied.

4.8 Show that if the input impedance to an amplifier arises from a pure capacitance C, then the effect of shunt-input negative feedback is to multiply C by the factor $(1 + A\beta)$. Further, show that the effect of series-input negative feedback is to cause C to be divided by the factor $(1 + A\beta)$.

4.9 A differential amplifier has a differential mode gain of A_{dm}. Show how series-parallel feedback can be applied using two resistors R_1 and R_2. Derive an expression for voltage gain after feedback is applied.

4.10 With suitable choice of turns ratio a transformer can be used to sample a current or voltage and provide any desired fraction at the transformer secondary. Show how series-voltage and shunt-current feedback can be applied to an amplifier by using a transformer for the feedback block.

Transformers were often used in the early days of feedback because the secondary can be reversed to provide negative β. This meant that only one active device, rather than two was needed for the forward amplifier. Transformers are used less often now because they are relatively expensive compared with active devices.

More about Feedback Amplifiers 5

Objectives

- ☐ To describe the more technical aspects of feedback amplifiers; loading effects and frequency-response effects.
- ☐ To analyse the loading effects that can occur between the forward amplifier and the feedback block.
- ☐ To extend this analysis method to account for loading effects on feedback amplifier behaviour of the self impedance of the signal source and also of the load impedance.
- ☐ To explain that negative feedback in general increases the bandwidth of an amplifier.
- ☐ To describe the significance of gain bandwidth product.
- ☐ To explain that when feedback is applied to an amplifier which is intended to be negative it can actually be positive at some frequencies. This can lead to unwanted peaking of the amplifier frequency response or even oscillations.
- ☐ To describe the conditions which lead to instability and conditional stability with reference to the Nyquist plot.

Loading Effects between the Forward Amplifier and the Feedback Block

In the discussion of feedback circuits in the previous chapter reference has been made to possible loading effects between the forward amplifier and the feedback block. If these are present and have a significant effect, the simple theory previously described in Chapter 4 may not predict the amplifier performance with sufficient accuracy. The modifications to the analysis methods are now discussed. A loading effect at the amplifier *output* by the feedback block *input* can occur because

(i) for *voltage-output* feedback, in sampling the output voltage the feedback block draws a finite current from the forward amplifier. As will be shown, this may cause a significant lowering of the forward gain A.

Ideally no current is drawn from the feedback amplifier.

(ii) for *current-output* feedback, in sampling the output current the feedback block develops a finite voltage drop which tends to reduce the output current and again may cause a significant lowering of the forward gain A.

Ideally no voltage is developed.

A loading effect at the feedback block *output* by the amplifier *input* can occur because

(i) for *shunt-input* feedback the feedback signal i_f is affected by the presence of a finite input voltage v_i (see Fig. 4.2b). This input voltage is developed across the closed-loop amplifier input impedance by the main input signal i_i. The voltage v_i also appears across the output terminals of the feedback block thus lowering i_f and hence the effective value of the feedback fraction β.

In effect the current coupling factor at the output of the feedback block is less than unity.

(ii) for *series-input* feedback the feedback signal v_f is affected by the presence of a finite current circulating round the input loop of the amplifier (see Fig. 4.2a).

That is, the voltage coupling factor is less than unity.

This current also passes through the output of the feedback block and reduces the output voltage of the feedback block thus lowering the effective value of feedback fraction β.

The changes in A and β caused by these loading effects do not affect the validity of the theory; for example, the fundamental feedback expression $A_f = A/(1 + A\beta)$. However, the values to be used for A and β are no longer given simply by considering the two parts (viz. forward amplifier and feedback block) in isolation. In some cases these loading effects cannot be neglected and an analysis method is used to assess them.

Such a method can also be useful where the loading effects are negligible but verification of this is needed.

In the method to be described two basic assumptions made about the feedback amplifier are as follows:

(i) *The forward amplifier is unilateral.* That is, signals flow through the amplifier from input to output, but not in the reverse direction.

(ii) *The feedback block is unilateral.* That is, signals pass from the point where the amplifier output signal is sampled, through the feedback block and are mixed with the main input signal, but do not pass along this path in the opposite direction.

These are reasonable assumptions to make. Practical forward amplifiers generally have little or no measurable reverse transmission, especially where the amplifier comprises several amplifying stages. Also, any small signal which does pass through the attenuating feedback block in the reverse direction can generally be neglected at the output when compared with the relatively much larger signal emanating from the forward path amplifier.

To develop the method, consider a particular feedback configuration, say the shunt-voltage feedback arrangement (Fig. 5.1). The forward amplifier is repre-

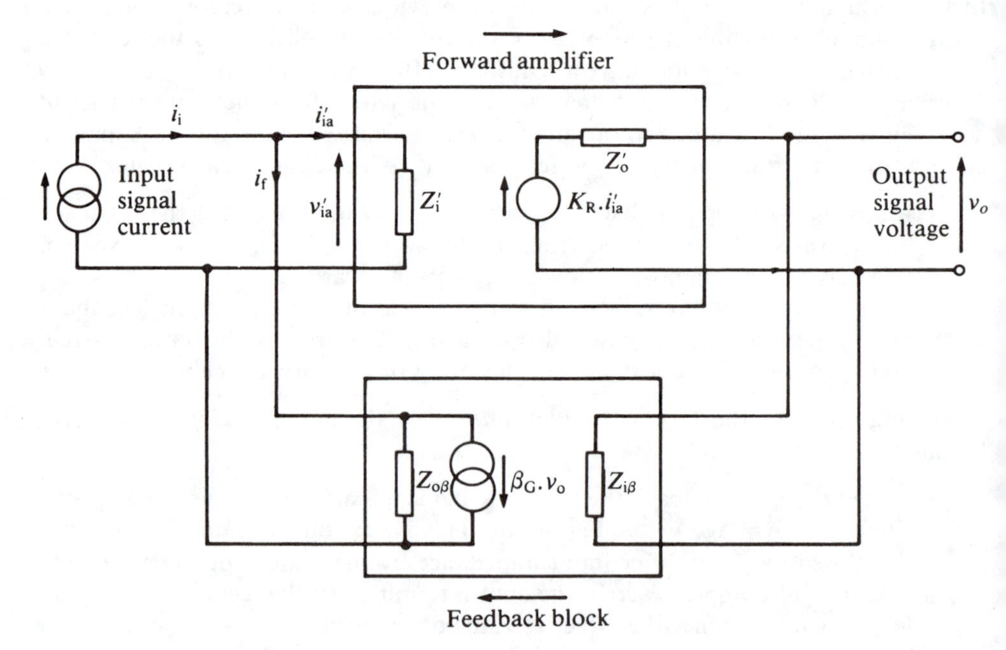

Fig. 5.1 Shunt-voltage arrangement used for development of a method for analysing loading effects.

sented by a general unilateral model comprising input and output impedances, Z'_i and Z'_o, and appropriate gain constant K_R. The feedback block samples the output voltage and returns this as a current signal. The feedback block also is represented by a general unilateral model. Ideally, it should behave as a voltage-controlled current source. However, it is necessary to include input and output impedances, $Z_{i\beta}$ and $Z_{o\beta}$ since their effect is to be considered. The feedback block in sampling the forward amplifier output voltage draws a small current and the feedback block input impedance $Z_{i\beta}$ is included to account for this. The output of the feedback block is connected across the input connections of the forward amplifier and acts as a source of current. Any variation in this current owing to loading effects occurs because the feedback block output does not act as an ideal current generator. Therefore, it is modelled by means of a current generator in parallel with a self impedance $Z_{o\beta}$.

The prime is used to distinguish the input and output impedances of the forward amplifier, z'_i and z'_o, from those of feedback circuit on open-loop, z_i and z_o.

This is a Norton equivalent circuit.

Defer, for a moment, the question of how to find $Z_{i\beta}$ and $Z_{o\beta}$ for a given practical circuit and first find how to analyse the arrangement shown in Fig. 5.1.

At first sight the circuit might seem quite complicated. However, the problem of analysing the circuit is made considerably easier if the $Z_{o\beta}$ and $Z_{i\beta}$ components are moved out of the feedback block and into the forward amplifier as shown in Fig. 5.2. This is done by imagining that the ends of the impedances $Z_{o\beta}$ and $Z_{i\beta}$ are slid along the conductors until the new positions are reached. Therefore, although the boundaries of the forward amplifier and feedback block have been redefined, the new circuit arrangement is electrically identical in overall behaviour to the original. The new arrangement is very much simpler to analyse because the new feedback block does not contain input or output impedances and it behaves as an ideal controlled source having *no associated loading effects*. The original loading

This is a conceptual move not an actual one. Circuit conditions are not changed because the voltage is everywhere the same on each conductor.

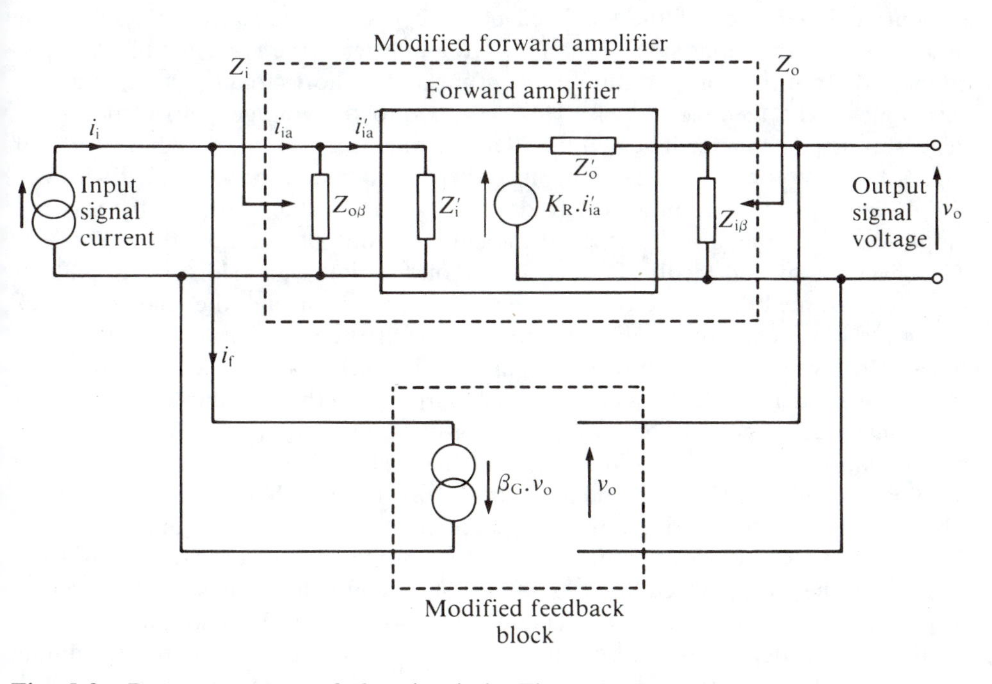

Fig. 5.2 Rearrangement of the circuit in Fig. 5.1 to move $Z_{o\beta}$ and $Z_{i\beta}$ into the forward amplifier block.

effects on the overall circuit behaviour are still there of course but their causes are now incorporated in the new representation of the forward amplifier.

For Fig. 5.2 the following open-loop quantities can be written down simply by inspection:

$$\text{Open-loop forward amplification } A = \text{Input current coupling factor} \times K_R \times \text{Output voltage coupling factor}$$

$$= \frac{Z_{o\beta}}{Z_{o\beta} + Z_i'} \cdot K_R \cdot \frac{Z_{i\beta}}{Z_{i\beta} + Z_o'}$$

and feedback fraction, $\beta = \beta_G$

Open-loop input impedance $Z_i = Z_{o\beta}||Z_i'$

Open-loop output impedance $Z_o = Z_{i\beta}||Z_o'$

Output impedance can be found by a new Thevenin equivalent circuit for the amplifier output, as shown in the figure below. The output impedance is the impedance looking back into the amplifier with the generator ($K_R.i_{ia}'$) set at zero.

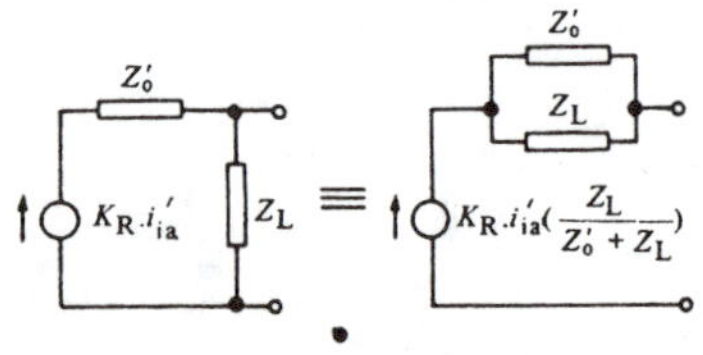

The feedback block is assumed to have no signal transmission in the reverse direction and no generators are required to represent this. Therefore the input to the feedback block only contains the impedance $z_{i\beta}$.

When these open-loop values are used in the expression for closed-loop quantities the values obtained are identical to those for the original circuit (Fig. 5.1). Therefore, provided the input and output impedances $Z_{i\beta}$ and $Z_{o\beta}$ of the feedback block are known, the above procedure provides an easy way to analyse a feedback circuit taking account of loading effects.

Now consider the problem deferred earlier. How, for a particular circuit, the input and output impedances of the feedback block, $Z_{i\beta}$ and $Z_{o\beta}$, as shown in Fig. 5.1, can be obtained. Observe that if the generator $\beta_G \cdot v_o$ in the feedback block is somehow made equal to zero then the whole circuit (forward amplifier and feedback block) is the same as the modified forward amplifier of Fig. 5.2. Hence, the output impedance of the feedback block $Z_{o\beta}$ is automatically included in the desired way at the input side of the forward amplifier if the feedback block is disconnected from the forward amplifier output and a short-circuit is applied across the input of the feedback block. This sets to zero the voltage-controlled current generator inside the feedback block. This is illustrated at the left hand side of Fig. 5.3b. Now for $Z_{i\beta}$, in Fig. 5.1. Since there is no generator associated with $Z_{i\beta}$, simply connecting the input of the feedback block across the output of the forward amplifier inserts the impedance $Z_{i\beta}$ at the amplifier output as required for Fig. 5.1. This gives the second feedback block shown on the right-hand side of Fig. 5.3b. It has been assumed that there is no reverse transmission through the feedback block and there is no need to specify the connections at the other end of the feedback block. However, in case, in practice, some small reverse transmission does occur, it is advisable to terminate the feedback block output with the impedance which it *sees* in normal working conditions when looking into the comparator. For the present case of shunt-input feedback voltage v_{ia} is small. Hence, the feedback block input terminals should be short-circuited as shown on the right-hand side of Fig. 5.3b. The circuit illustrated in Fig. 5.3b (and all the other circuits in the figure) shows two feedback blocks whereas in the feedback circuit proper there is only one. This is not so strange when it is realised that it is a fictitious circuit! The purpose of the circuit is to provide a convenient way to calculate the forward gain of the amplifier.

Having considered just one of the four possible configurations, the remaining configurations can be treated in the same way by modifying the forward amplifier to include the input and output impedances of the feedback block. The way to do this

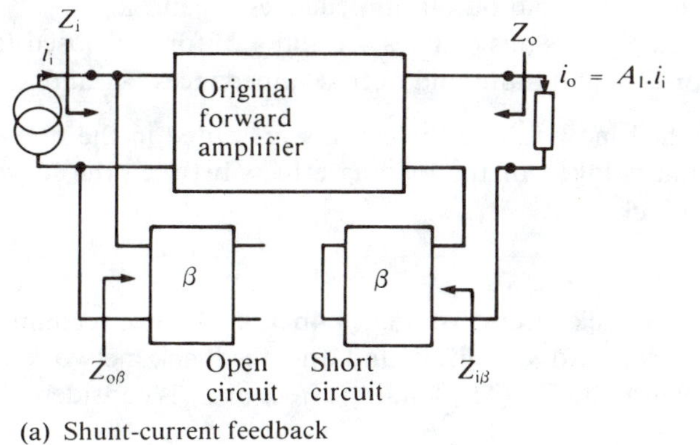

(a) Shunt-current feedback

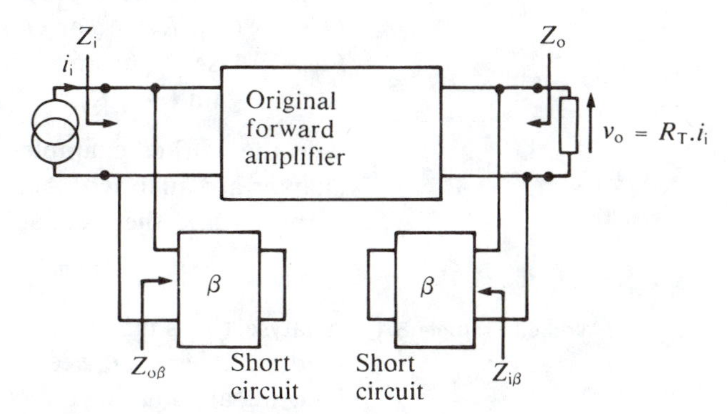

(b) Shunt-voltage feedback

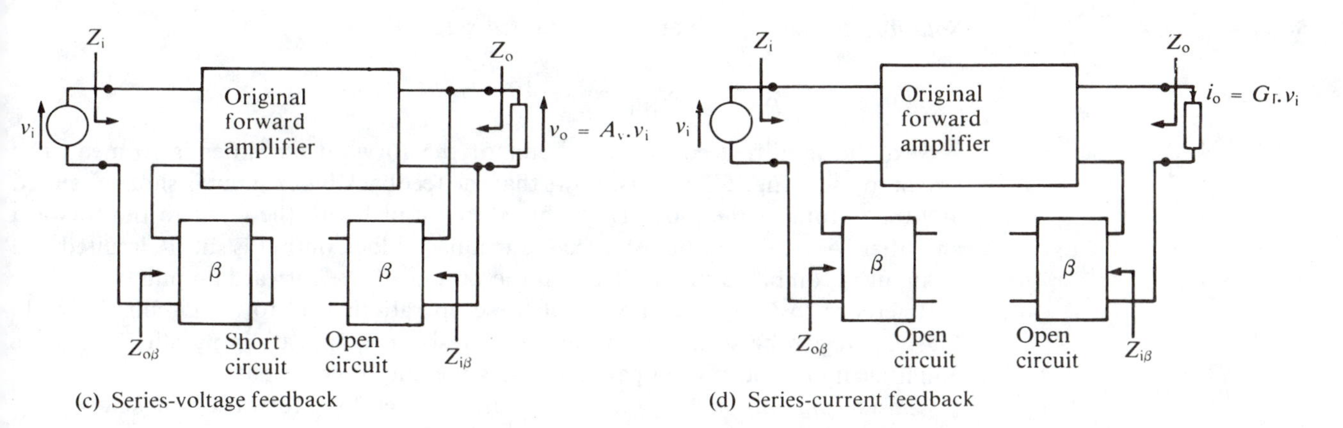

(c) Series-voltage feedback

(d) Series-current feedback

Fig. 5.3 Formation of the modified forward amplifier.

is shown in the other parts of Fig. 5.3. It is not necessary to derive these other conditions (although of course it is possible to do so, and you may wish to check them for yourself) because from the experience of analysing other properties of feedback circuits it is known that they must conform to a pattern. Each of the four circuits contains two feedback blocks. One is connected to the forward amplifier input in parallel or series in accordance with the type of feedback configuration, i.e. parallel-input or series-input respectively, the other is connected to the output of the forward amplifier in a similar manner. The remaining ends of the feedback blocks are terminated with short-circuits or open-circuits. The left hand block is terminated in a way which breaks the feedback path, i.e. a short circuit for voltage-output feedback and an open-circuit for current-output feedback. The right-hand feedback block is terminated with a short-circuit if the feedback is shunt-input and an open-circuit if the feedback is series-input.

The feedback configurations in Fig. 5.3 are in the same sequence as in previous diagrams, e.g. Fig. 4.1

Before considering some worked examples the steps in the method are summarized:

Step 1 Set the feedback fraction β equal to the ideal unloaded transfer ratio of the feedback block.

A more mathematical treatment of this method is to be found in Gray, P.E. and Searle, C.L. *Electronic Principles* (Wiley, 1967). Also see Millman, J. *Microelectronics* (McGraw-Hill, 1979).

Step 2 Using the appropriate circuit form in Fig. 5.3, calculate the open-loop gain (A_I, R_T, A_V or G_T) and input and output impedances, Z_i and Z_o.

Step 3 Use the basic feedback expressions (Tables 4.1 and 4.2) for the closed-loop gain (A_{If}, R_{Tf}, A_{Vf} or G_{Tf}) and input and output impedances, Z_{if} and Z_{of}.

In the worked examples which now follow the circuits presented in the previous chapter are analysed. Account is taken of the loading effects between the forward amplifier and the feedback block.

Worked Example 5.1 Analyse the shunt-voltage feedback circuit of Fig. 4.4b taking into account the loading effects between the forward amplifier and the feedback network, but ignoring any loading effect caused by Z_L. (The loading effect of Z_L is considered in a later section.)

Step 1.

Solution: Following the above mentioned steps, first

$$\beta_G = -\frac{1}{R} = -\frac{1}{10\ \text{k}\Omega} = -10^{-4}\ \text{S} \tag{5.1}$$

Next the modified equivalent circuit of the forward amplifier is formed. The circuit to use is Fig. 5.3b. This shows that the feedback block input is short-circuited and the output of the block is combined in parallel with the input to the forward amplifier. Also, the output of a second feedback block output is short-circuited and its input is combined in parallel with the output of the forward amplifier.

Reference to Fig. 4.4a shows that these operations lead to the circuit shown in Fig. 5.4. It can be seen that the feedback resistor R appears in parallel across the amplifier input and also in parallel across the amplifier output.

Step 2.

Considering only small signal quantities, the open-loop forward gain is now given by

$$R_T = \frac{v_o}{i_{ia}} = \frac{i_b}{i_{ia}} \cdot \frac{(-i_c)}{i_b} \cdot \frac{v_o}{(-i_c)}$$

$$= \text{Input current coupling factor} \times -h_{fe} \times \text{Effective collector resistance}$$

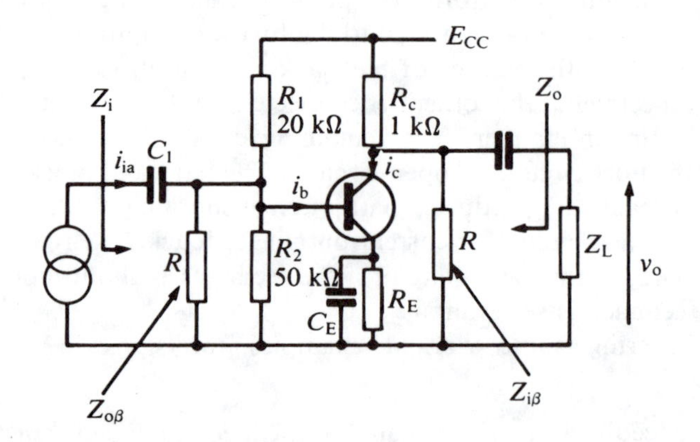

Fig. 5.4 Modified forward amplifier. (For the BJT, h_{ic} = 1 kΩ, h_{fe} = 100.)

$$= \frac{R_1||R_2||R}{(R_1||R_2||R) + h_{ie}} \cdot (-h_{fe}) \cdot R_c||R$$

For a.c. signals the voltage supply E_{CC} is effectively short-circuited to the common rail, thus placing R_c and R in parallel.

$$= \frac{20||5||10}{20||5||10 + 1} \cdot (-100) \cdot 1||10 \text{ k}\Omega$$

$$= .741 \times (-100) \times 0.909 \text{ k}\Omega$$

So, $R_T = -67.3 \text{ k}\Omega$. (5.2)

This may be compared with the value of -80 kΩ calculated in Chapter 4 ignoring this loading effect.

Therefore

$$(1 + R_T\beta_G) = 1 + (-67.3 \times 10^3)(-10^{-4}) = 7.73 \quad (5.3)$$

It may be noted that the previous value was 9.

Disregarding the reactances of the capacitances, as previously, the input and output impedances of the modified amplifier are also obtained from Fig. 5.4, as follows:

$$Z_i = R_1||R_2||R||h_{ie} = 20||5||10||1 \text{ k}\Omega = 741 \ \Omega \quad \text{(previously 800 }\Omega) \quad (5.4)$$

$$Z_o = R_c||R = 1||10 \text{ k}\Omega = 909 \ \Omega \quad \text{(previously 1000 }\Omega) \quad (5.5)$$

Step 3.

The closed-loop parameters can now be found using the values calculated and the expressions in Tables 4.1 and 4.2.

Closed-loop gain, $R_{Tf} = \dfrac{R_T}{1 + R_T\beta_G}$

$$= \frac{-67.3}{7.73} \text{ k}\Omega$$

$$= -8.71 \text{ k}\Omega \quad \text{(previously } -8.89 \text{ k}\Omega)$$

Closed-loop input resistance, $r_{if} = \dfrac{741}{7.73} \ \Omega$

$$= 96 \ \Omega \quad \text{(previously 89 }\Omega)$$

Closed-loop output resistance, $r_{of} = \dfrac{909}{7.73} \ \Omega$

$$= 118 \ \Omega \quad \text{(previously 111 }\Omega)$$

It is interesting to observe that in the above example although the loading produces a noticeable effect on open-loop gain, as shown by Equation 5.2, the values of the closed-loop quantities are less affected. This illustrates the fact that negative feedback circuits are relatively unaffected by loading effects as well as by variations in the parameters of the forward amplifier.

Worked Example 5.2

Analyse the series-voltage feedback circuit of Fig. 4.5b. Take into account loading effects between the forward amplifier and the feedback network.

Solution: The feedback block in this circuit is a potential divider formed from resistors R_1 and R_2. The unloaded feedback fraction is thus given by (see Equation 4.24)

Step 1.

$$\beta_V = \frac{R_2}{R_1 + R_2} = \frac{2\ \text{k}\Omega}{10\ \text{k}\Omega + 2\ \text{k}\Omega} = \frac{1}{6} \tag{5.6}$$

The modified forward amplifier is obtained this time using Fig. 5.3c. This shows that a feedback block is connected in parallel at the output of the forward amplifier with the other end of the feedback block open-circuited. The result of this step is to place the series combination of R_1 and R_2 across the amplifier output as shown at the right-hand side of Fig. 5.5. In addition, a second feedback block is connected in series with the amplifier input while the other end of this block is short-circuited. Fig. 4.5a shows that this operation has the effect of placing the combination of $R_1 \| R_2$ in series at the lower input terminal of the forward amplifier. This lower terminal represents the base b_2 in Fig. 4.5b. Therefore, in Fig. 5.5 we put the parallel combination $R_1 \| R_2$ between base b_2 and the lower side of the signal input v_{ia}.

Step 2.

It should be noted that the input resistance between the two bases b_1 and b_2 given by $2h_{ie}$ is now represented in Fig. 5.5 by r_i' and is no longer denoted by r_i (see Equation 4.26). The unprimed r_i is used throughout for the terminal input resistance seen by the signal source.

Next the open-loop gain is calculated. First note that $R_1 \| R_2$ and r_i' of the long-tailed pair give rise to a voltage coupling effect. Also, note that the effective collector resistance of transistor T_2 is $R_c \| (R_1$ in series with $R_2)$. Hence

$$A_V = \frac{v_o}{v_{ia}} = \frac{v_{ia}}{v_{ia}} \cdot \frac{v_o}{v_{ia}'} = \frac{r_i'}{r_i' + (R_1 \| R_2)} \cdot \frac{v_o}{v_{ia}}$$

Now the ratio v_o / v_{ia}' is the differential-mode gain and using Equation 4.25 with R_c replaced by its effective value gives for A_V

$$A_V = \frac{r_i'}{r_i' + (R_1 \| R_2)} \cdot \frac{1}{2} h_{fe} \cdot \frac{R_c \| (R_1 + R_2)}{h_{ie}}$$

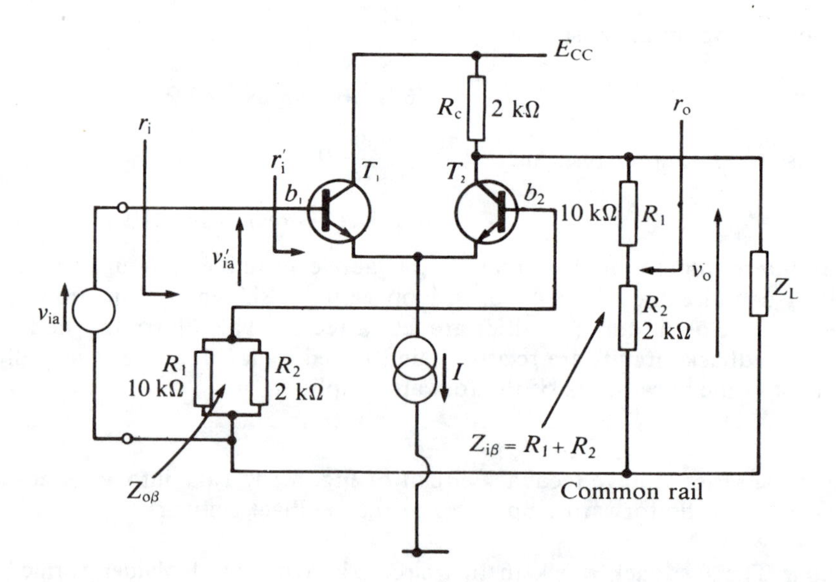

Fig. 5.5 Modified forward amplifier. (For the BJTs T_1 and T_2, $h_{ic} = 1\ \text{k}\Omega$, $h_{fe} = 100$.)

Substituting $r_i' = 2\,h_{ie}$ and numerical values,

$$A_V = \frac{2\text{ k}\Omega}{2\text{ k}\Omega + (10\text{ k}\Omega||2\text{ k}\Omega)} \cdot \left(\frac{1}{2} \cdot 100\right) \cdot \frac{2\text{ k}\Omega||(10\text{ k}\Omega + 2\text{ k}\Omega)}{1\text{ k}\Omega}$$

$$= 0.545 \times 50 \times 1.714 = 46.7 \quad \text{(previously 100)} \qquad (5.7)$$

The calculated value of A is now very different from the value obtained when loading effects were ignored.

Hence

$$(1 + A_V \cdot \beta_V) = 1 + 46.7 \times \tfrac{1}{6} = 8.79 \quad \text{(previously 17.7)} \qquad (5.8)$$

From Fig. 5.5 and Equation 4.26 the value of open-loop input resistance as seen by the signal source is

$$r_i = r_i' + R_1||R_2 = 2 \times 1\text{ k}\Omega + 10\text{ k}\Omega||2\text{ k}\Omega$$

$$= 3.67\text{ k}\Omega \quad \text{(previously 2 k}\Omega\text{)} \qquad (5.9)$$

Also, the value of the open-loop output resistance is equal to the effective collector resistance (see Equation 4.27) and so

$$r_o = R_c||(R_1 + R_2) \qquad (5.10)$$

$$= 2\text{ k}\Omega||(10\text{ k}\Omega + 2\text{ k}\Omega) = 1.71\text{ k}\Omega \quad \text{(previously 2 k}\Omega\text{)}$$

The expressions in Tables 4.1 and 4.2 and the above quantities provide the closed loop quantities as follows

Step 3.

$$\text{Closed-loop gain, } A_{vf} = \frac{A_V}{1 + A_V\beta_V} = \frac{46.7}{8.79} = 5.32 \text{ (previously 5.66)} \qquad (5.11)$$

$$\text{Closed-loop input resistance, } r_{if} = r_i(1 + A_V\beta_V)$$

$$= 3.67 \times 8.79\text{ k}\Omega$$

$$= 32.2\text{ k}\Omega \text{ (previously 35.3 k}\Omega\text{)} \qquad (5.12)$$

$$\text{Closed-loop output resistance } r_{of} = \frac{r_o}{1 + A_V\beta_V} = \frac{1.71}{8.79}\text{ k}\Omega$$

$$= 195\ \Omega \text{ (previously 113 }\Omega\text{)} \qquad (5.13)$$

In this circuit loading has a more marked effect. The forward gain is reduced to less than a half of its unloaded value, as Equation 5.7 shows. It can be seen from the calculations that this is principally owing to the loading effect at the input side of the amplifier, because the r_i' and $R_1||R_2$ produce the quite low voltage coupling factor of 0.545. This can be improved by reducing $R_1||R_2$. The values of R_1 and R_2 can be varied without affecting β_V provided the *relative* values of R_1 and R_2 are kept in the ratio 2 kΩ:10 kΩ since they determine β_V. However, it should be noted that a reduction of R_1 and R_2 to reduce loading at the input side is at the expense of worse loading at the output side.

Although this does not effect the closed-loop gain very much, the feedback factor $(1 + A\beta)$ is significantly changed so that benefits such as reduced sensitivity are affected.

The designer can explore the choice of values to try to obtain the best mix of closed-loop amplifier characteristics (including input and output resistances) for the intended application.

Worked Example 5.3

Analyse the series-current transistor feedback circuit of Fig. 4.6b. Take into account loading effects between the forward amplifier and the feedback network.

Solution: From the discussion of this circuit in Chapter 4 it is known that the unloaded feedback fraction of this circuit is

Step 1.

$$\beta_R = -R$$

where R is the resistance in the emitter path. To form the modified open-loop amplifier circuit allowing for loading, Fig. 5.3d is used. This shows that an open-circuited feedback block (as shown in Fig. 4.6a) is inserted in series with the input loop and a second open circuited block in series with the output loop. The result of this operation is shown in Fig. 5.6.

Step 2.

The open-loop forward gain is required:

$$G_T = \frac{i_o}{v_{ia}} \tag{5.14}$$

From the previous analysis of this circuit in Chapter 4 the relationship between i_o and the base-emitter voltage (there called v_{ia} but here labelled v'_{ia} since v_{ia} is reserved for overall input voltage to the forward amplifier) is given by

$$\frac{i_o}{v'_{ia}} = \frac{-h_{fe}}{h_{ie}} \tag{5.15}$$

It can be seen from Fig. 5.6 that v_{ia} and v'_{ia} are related by the voltage coupling factor formed by the input resistance to the transistor, $r'_i = h_{ie}$, and R. Thus

$$v'_{ia} = \frac{r'_i}{r'_i + R} \cdot v_{ia} = \frac{h_{ie}}{h_{ie} + R} \cdot v_{ia} \tag{5.16}$$

Substituting v'_{ia} from Equation 5.16 into Equation 5.15 gives the required expression for G_T in Equation 5.14.

Check this yourself.

$$G_T = -\frac{h_{fe}}{h_{ie} + R} \tag{5.17}$$

The open-loop input resistance is given by inspection of Fig. 5.6 as follows

Resistor R is effectively in series with the input resistance to the amplifier r'_i.

$$r_i = r'_i + R = h_{ie} + R \tag{5.18}$$

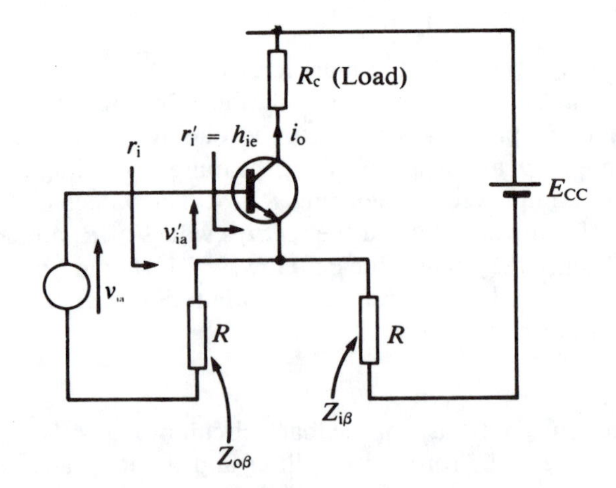

Fig. 5.6 Modified forward amplifier.

The open-loop output resistance r_o as seen by the load resistance in the collector is infinitely large since the load is in series with the infinitely large collector resistance of the transistor.

The above quantities can now be substituted in the closed-loop expressions of Tables 4.1 and 4.2 to give the following:

Step 3.

$$\text{Closed-loop gain, } G_{Tf} = \frac{G_T}{1 + G_T\beta_R} = \frac{\dfrac{-h_{fe}}{h_{ie} + R}}{1 + \dfrac{-h_{fe}}{h_{ie} + R}(-R)}$$

$$= \frac{-h_{fe}}{h_{ie} + R + h_{fe}R} \tag{5.19}$$

Therefore

$$G_{Tf} = \frac{-h_{fe}}{h_{ie} + (1 + h_{fe})R} \tag{5.20}$$

$$\text{Closed-loop input resistance, } r_{if} = r_i(1 + G_T \cdot \beta_R)$$

$$= (h_{ie} + R)\left(1 + \frac{-h_{fe}}{h_{ie} + R}(-R)\right)$$

$$= h_{ie} + R + h_{fe}R$$

Therefore

$$r_{if} = h_{ie} + (1 + h_{fe})R \tag{5.21}$$

$$\text{Closed-loop output resistance, } r_{of} = r_o(1 + G_T \cdot \beta_B)$$

$$= \infty \tag{5.22}$$

since $r_o \to \infty$

It should be recalled that when this circuit was analysed without considering loading effects small discrepancies occurred in relations to the expressions for G_{Tf} and r_{if}. A comparison of the above Equations 5.20 and 5.21 with the exact expressions obtained from transistor circuit analysis methods, Equations 4.29 and 4.30, shows that the discrepancy has now been removed.

Exercise 5.1

Evaluate, with and without loading effects, the values of G_{Tf} and r_{if} assuming the typical values $h_{fe} = 100$, $h_{ie} = 1$ kΩ, $R = 1$ kΩ.
[Answer: with loading effects G_{Tf}, r_{if} are -9.80×10^{-4} S, 102 kΩ; without loading effects, -9.90×10^{-4} S, 101 kΩ respectively.]

Worked Example 5.4

Analyse the circuit of Fig. 4.7b given that $R_1 = 10$ kΩ, $R_2 = 2$ kΩ and $R_{c1} = 5$ kΩ. Assume the transistors are identical and the only significant parameters are $h_{ie} = 1$ kΩ, $h_{fe} = 100$. As in the previous example assume that the transistor collector resistances are infinite.

Solution: This is a shunt-current feedback configuration, for which the unloaded feedback fraction (see Equation 4.36) is given by

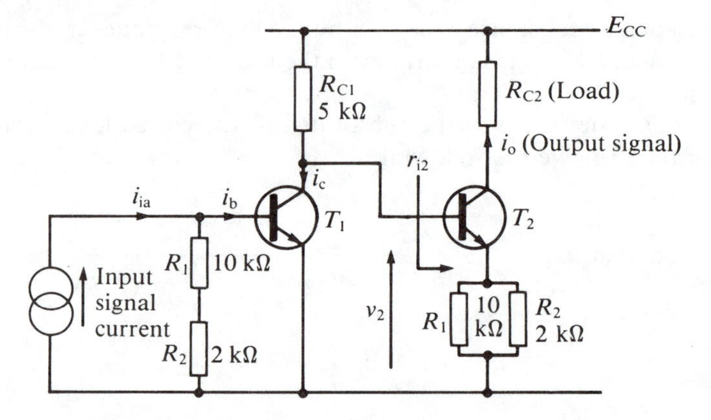

Fig. 5.7 Modified forward amplifier. (For the BJTs T_1 and T_2, $h_{ie} = 1$ kΩ, $h_{fe} = 100$.)

Step 1.

$$\beta_I = \frac{R_2}{R_1 + R_2} = \frac{2}{10 + 2} = \frac{1}{6} \tag{5.23}$$

Step 2.

The modified forward gain amplifier is constructed according to Fig. 5.3a. Referring to Fig. 4.7a, to give the right-hand block in Fig. 5.3a representing the loading of the feedback block on the amplifier output, the procedure is to connect the left hand terminal of R_1 to the common rail. This places the combination $R_1||R_2$ in the emitter lead of T_2 as shown in Fig. 5.7. The loading effect at the input of the amplifier is obtained by open-circuiting a second feedback block at R_2 which results in the series combination $(R_1 + R_2)$ being placed across the amplifier input as also shown in Fig. 5.7.

For this circuit it is required to calculate the open-loop forward current amplification,

Incremental (i.e. *small signal*) quantities are assumed.

$$A_I = \frac{i_o}{i_{ia}} \tag{5.24}$$

Current i_{ia} provides i_b which results in i_c and then v_2. This voltage acts on the second transistor to give the output signal i_o.

Following the input signal, i_{ia}, through the circuit,

$$i_b = \text{Input-current coupling factor} \times i_{ia}$$

$$= \frac{(R_1 + R_2)}{(R_1 + R_2) + h_{ie}} \cdot i_{ia}$$

$$= \frac{10 + 2}{10 + 2 + 1} \cdot i_{ia}$$

$$= 0.923 \cdot i_{ia} \tag{5.25}$$

The base current i_b is multiplied by h_{fe} to give the collector current i_c of the first transistor. In turn i_c causes a collector voltage v_2 across the equivalent collector load of the first transistor, i.e. $R_{c1}||r_{i2}$. Thus

$$v_2 = (-h_{fe})(R_{c1}||r_{i2})\cdot i_b \tag{5.26}$$

An example of multiple feedback.

Here r_{i2} is the input resistance to the second transistor. The effect of $R_1||R_2$ in the emitter path of the second transistor is to apply local series-current feedback of the type analysed in Worked Example 5.3. Using the results from that worked example

$$r_{i2} = h_{ie} + (1 + h_{fe})(R_1||R_2) \quad (5.27)$$

$$= 1\ \text{k}\Omega + (1 + 100)(10\ \text{k}\Omega||2\ \text{k}\Omega)$$

$$= 169\ \text{k}\Omega$$

Thus from Equation 5.26

$$v_2 = (-100)(5\ \text{k}\Omega||169.3\ \text{k}\Omega) \cdot i_b = -4.86 \times 10^5 \times i_b \quad (5.28)$$

Continuing to follow the signal through the circuit, v_2 causes i_o as given by G_{Tf} in Equation 5.20 with $R_1||R_2$, substituting for R,

$$\frac{i_o}{v_2} = \frac{-h_{fe}}{h_{ie} + (1 + h_{fe})(R_1||R_2)}$$

$$= \frac{-100}{1\ \text{k}\Omega + (1 + 100)(10\ \text{k}\Omega||2\ \text{k}\Omega)}$$

$$= -5.91 \times 10^{-4}\ \text{S} \quad (5.29)$$

These equations are put together to give the open-loop current gain,

$$A_I = \frac{i_o}{i_{ia}} = \frac{i_b}{i_{ia}} \cdot \frac{v_2}{i_b} \cdot \frac{i_o}{v_2}$$

$$= 0.923 \times (-4.86 \times 10^5) \times (-5.91 \times 10^{-4}) = 265 \quad (5.30)$$

The other quantities required are the open-loop input and output resistances. By inspection of Fig. 5.7,

$$\text{Open-loop input resistance } r_i = (R_1 + R_2)||h_{ie}$$

$$= (10\ \text{k}\Omega + 2\ \text{k}\Omega)||1\ \text{k}\Omega = 923\ \Omega \quad (5.31)$$

The open-loop output resistance r_o as seen by the load resistance R_{C2} is infinitely large since the load is in series with the infinitely large collector resistance of the transistor.

The expressions of Tables 4.1 and 4.2 can now be used to calculate the closed-loop quantities. The calculations are as follows:

$$\text{Closed-loop current gain, } A_{If} = \frac{A_I}{1 + A_I\beta_I}$$

$$= \frac{265}{1 + 265 \times \frac{1}{6}}$$

$$= 5.87 \quad (5.32)$$

$$\text{Closed-loop input resistance, } r_{if} = \frac{r_i}{1 + A_I \cdot \beta_I}$$

$$= \frac{923}{1 + 265 \times \frac{1}{6}}$$

$$= 20.4\ \Omega$$

$$\text{Closed-loop output resistance, } r_{of} = r_o(1 + A_I\beta_I)$$
$$= \infty \quad (\text{since } r_o \rightarrow \infty)$$

Now try the following exercise.

Exercise 5.2 Work out the closed-loop gain, and closed-loop input and output resistances for the feedback amplifier in Fig. 4.7b, disregarding loading effects and compare the results with those obtained immediately above.
[Answer: disregarding loading a much higher value of open-loop gain, $A_I = 8000$, is obtained, which leads to closed-loop values, $A_{If} = 5.996$, $r_{if} =. .794\ \Omega$, $r_{of} = \infty$.]

Effect on Feedback Amplifier of Source and Load Impedances

The previous section has dealt at some length with the loading effects that can occur between the feed-forward amplifier and the feedback block. Loading effects can also occur because of the self impedance Z_g of the signal source and the load impedance Z_L connected at the feedback amplifier output.

As is the case with loading associated with the feedback block the present loading effects are in many cases small enough to be neglected. However, in other cases it is important to include them in calculations. This section shows how to do this.

It is appreciated that the value of the load impedance tends to affect the gain of the forward amplifier. This in turn affects the closed-loop performance of the circuit. As is seen, the signal source self impedance acts as a load on the feedback signal at the amplifier input and this also affects the behaviour of the feedback circuit.

This is similar in concept to the method previously described for loading effects between forward amplifier and feedback block.

One way to analyse these effects is to imagine that the source and load impedances are incorporated in the amplifier. To see how this is done consider one of the four feedback configurations. As before the shunt-voltage feedback arrangement is chosen (Fig. 5.1). The figure shows the input and output impedances, Z_i' and Z_o', of the forward amplifier and also those of the feedback block, $Z_{i\beta}$ and $Z_{o\beta}$. To examine the loading affects associated with the feedback block the impedances $Z_{i\beta}$ and $Z_{o\beta}$ are moved, as before, into the forward amplifier as shown in Fig. 5.2.

This can be confirmed by inspection of Fig. 4.1.

In addition, it is necessary to account for the effects of the self impedance of the main signal source Z_g, and the load impedance Z_L at the output of the feedback amplifier. These are added in parallel, respectively, with the input signal source and the amplifier output and can therefore be placed in parallel with the input and output impedance of the forward amplifier as shown in Fig. 5.8.

Redrawing the box does not affect the electrical behaviour of the circuit of course. The box boundary is not connected electrically to the circuit and exists only in our minds.

For the purposes of analysis Z_g and Z_L may be thought of as part of a new feedback circuit in which the new forward amplifier includes these elements as indicated by the broken lines in Fig. 5.8. This new feedback amplifier is fed by an ideal current signal source and has no external load impedance connected. Therefore once the new feedback amplifier circuit has been analysed loading effects are automatically included. Simplifications in the analysis can now be seen.

At the input side of the circuit Z_g and $Z_{o\beta}$ are connected in parallel and can be considered as a single impedance which reduces both the input signal J_g and the feedback current, $\beta_G \cdot v_o$, from entering the input to the forward amplifier. At the output side Z_L is in parallel with $Z_{i\beta}$ and can be considered as a single impedance

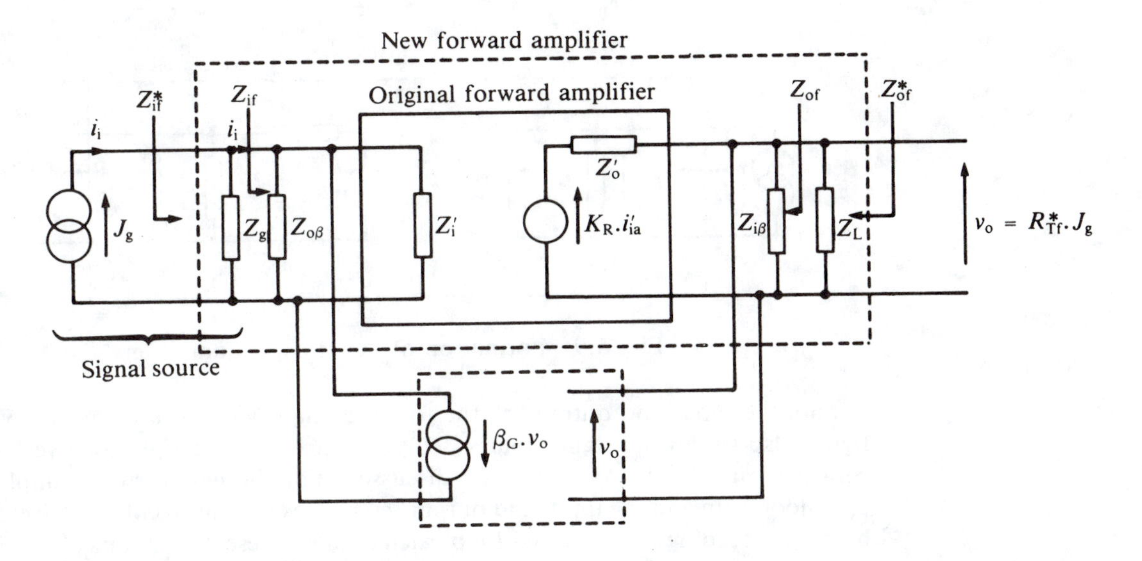

Fig. 5.8 Formation of new forward amplifier to take account of Z_g and Z_L loading.

which loads the forward amplifier. Hence, although the circuit at first glance may look more complicated, it is the same as in the previous section where only the loading effects between the feedback block and the forward amplifier were considered.

This means that the method described earlier to analyse feedback circuits with loading can also be used for the present case. An addition to that method is required to account for the input and output points of the new feedback circuit not being exactly those of the actual feedback circuit before the impedances Z_g and Z_L are incorporated.

The details of this are made clear in the following worked example.

Worked Example 5.5

A shunt-voltage feedback circuit is constructed using the arrangement shown in Fig. 4.4b and can be represented by the arrangement shown in Fig. 5.8. The forward amplifier has purely resistive input and output impedances of $r'_i = 1$ kΩ, $r'_o = 2$ kΩ and a trans-resistance gain constant of $K_R = -10^6$ Ω. The feedback resistance R is 10 kΩ. A 1 mA signal source J_g having a self resistance of $r_g = 2$ kΩ is applied at the input and a load resistance of $r_L = 5$ kΩ is connected at the feedback amplifier output. Calculate (a) the output voltage, (b) the input resistance seen by the actual signal source (including Z_g), (c) the output resistance seen by the load, and (d) the closed loop gain v_o/i_i of the amplifier circuit.

Solution: From previous discussion of this type of configuration the unloaded feedback fraction (Equation 5.1) for this circuit is given by

$$\beta_G = -\frac{1}{R} = -\frac{1}{10\ \text{k}\Omega} = -10^{-4}\ \text{S} \tag{5.33}$$

The next step is to construct the new forward amplifier. To do this the loading effect associated with the feedback block is accounted for using the procedure indicated by Fig. 5.3b. This results in the feedback resistance R appearing in parallel

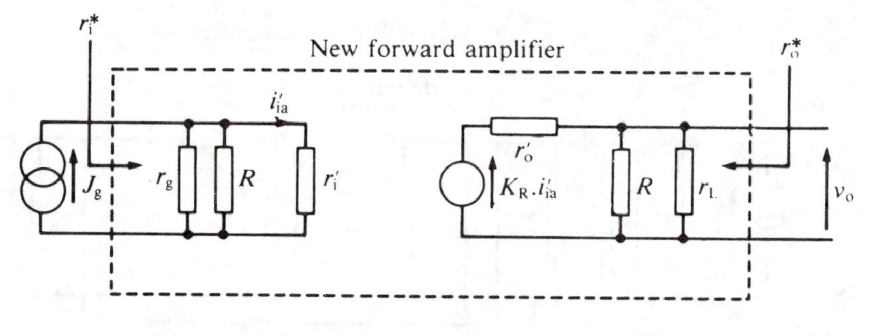

Fig. 5.9 Formation of new forward amplifier.

at both the input and outputs of the forward amplifier as shown in Fig. 5.9. This figure also shows the signal source self resistance r_g and output load resistance r_L connected in parallel as previously discussed. For the new forward amplifier the open-loop gain and the input and output resistances without feedback are calculated before proceeding to the closed-loop calculations. These open-loop calculations are as follows:

The asterisk* is used to denote quantities when resistance r_g and r_L are included in the forward amplifier.

$$\text{Open-loop gain, } R_T{}^* = \frac{v_o}{J_g}$$

$$= \begin{matrix}\text{Input current} \\ \text{coupling factor}\end{matrix} \times K_R \times \begin{matrix}\text{Output voltage} \\ \text{coupling factor}\end{matrix} \quad (5.34)$$

$$= \frac{(r_g \| R)}{(r_g \| R) + r'_i} \times K_R \times \frac{(R \| r_L)}{(R \| r_L) + r'_o}$$

$$= \frac{(2\|10)}{(2\|10) + 1} \cdot (-10^6) \cdot \frac{(10\|5)}{(10\|5) + 1} \text{ k}\Omega$$

So

$$R_T{}^* = 0.625 \times (-10^6) \times 0.454 = -284 \text{ k}\Omega \quad (5.35)$$

Next, by inspection of Fig. 5.9,

$$\text{Open-loop input resistance, } r_i{}^* = r_g \| R \| r'_i$$

$$= 2 \text{ k}\Omega \| 10 \text{ k}\Omega \| 1 \text{ k}\Omega = 625\ \Omega \quad (5.36)$$

The output impedance is obtained as before by setting the amplifier voltage generator to zero and looking into the circuit at the output terminals.

and

$$\text{Open-loop output resistance, } r_o{}^* = r'_o \| R \| r_L$$

$$= 4 \text{ k}\Omega \| 10 \text{ k}\Omega \| 5 \text{ k}\Omega = 1820\ \Omega \quad (5.37)$$

Using $\beta_G = -10^{-4}$ and the above open-loop values, the closed-loop quantities are now calculated using the expression in Tables 4.1 and 4.2, as follows:

$$\text{Closed-loop gain, } R_{Tf}{}^* = \frac{R_T{}^*}{1 + R_T{}^*\beta_G}$$

$$= \frac{-284 \text{ k}}{1 + (-284 \text{ k})(-10^{-4})}$$

$$= \frac{-284 \text{ k}}{1 + 28.4}$$

$$= -9.66 \text{ k}\Omega \quad (5.38)$$

$$\text{Closed-loop input resistance, } r_{if}^* = \frac{r_i^*}{1 + R_T^* \cdot \beta_G}$$

$$= \frac{625}{1 + 28.4} = 22.0\ \Omega \qquad (5.39)$$

$$\text{Closed-loop output resistance, } r_{of}^* = \frac{r_o^*}{1 + R_T^* \cdot \beta_G}$$

$$= \frac{1820}{1 + 28.4} = 61.8\ \Omega \qquad (5.40)$$

The parameters required in the question can now be calculated.

(a) *The output voltage.* Directly from Fig. 5.8 we have that

$$v_{out} = R_{Tf}^* \cdot J_g = (-9.66\ \text{k}\Omega)\,(1\ \text{mA}) = -9.66\ \text{V} \qquad (5.41)$$

The supply voltage E_{CC} must of course be sufficiently large to enable this output signal voltage to be developed without saturating the transistor.

(b) *The input resistance seen by the actual signal source.* Inspection of Fig. 5.8 shows that r_{if} is required (marked on the figure as impedance Z_{if}). The input resistance r_{if}^* is equal to r_g in parallel with r_{if}. Thus

$$r_{if}^* = r_g \| r_{if} = \frac{r_g \times r_{if}}{r_g + r_{if}} \qquad (5.42)$$

It is a simple matter to rearrange this expression to give

$$r_{if} = \frac{r_g \times r_{if}^*}{r_g - r_{if}^*} = \frac{2\ \text{k} \times 22.0}{2\ \text{k} - 22.0}\ \Omega = 22.2\ \Omega \qquad (5.43)$$

Another way to do this is to work in conductances. Define $g_{if}^* = 1/r_{if}^*$, $g_g = 1/r_g$, $g_{if} = 1/r_{if}$. Then input conductance seen by the signals source is $g_{if}^* = g_g + g_{if}$ (conductances in parallel are added). Therefore, $g_{if} = g_{if}^* - g_g$. Exercise: check this gives the same result for r_{if}.

(c) *The output resistance seen by the load.* In this case r_{of} is required (marked as impedance Z_{of} in Fig. 5.8) and is related to r_{of}^* by

$$r_{of}^* = r_{of} \| r_L = \frac{r_{of} \cdot r_L}{r_{of} + r_L}$$

Rearranging,

$$r_{of} = \frac{r_L \times r_{of}^*}{r_L - r_{of}^*} = \frac{5\ \text{k} \times 61.82}{5\ \text{k} - 61.82} = 62.6\ \Omega \qquad (5.44)$$

As above, another way is to use $g_{of} = g_{of}^* - g_L$, and $r_{of} = 1/g_{of}$.

(d) *The closed-loop gain of the amplifier circuit.* The closed-loop gain R_{Tf}^* calculated above in Equation 5.38 is not that of the actual amplifier but of a hypothetical amplifier circuit with r_g and r_L incorporated. The following is required

$$R_{Tf} = \frac{v_o}{i_i} \quad \text{rather than} \quad R_{Tf}^* = \frac{v_o}{J_g} \qquad (5.45)$$

A simple way to obtain R_{Tf} is found by considering the point in the circuit of Fig. 5.8 where the input signal i_i flows into the actual amplifier. The input resistance to the amplifier at this point is r_{if} (marked as Z_{if} on the figure). Hence, the current source J_g is coupled to the current i_i with a coupling factor determined by the two resistances r_g and r_{if}. Thus

$$i_i = \text{Current coupling factor} \times J_g = \frac{r_g}{r_g + r_{if}} \cdot J_g \qquad (5.46)$$

This equation and Equation 5.45 may be combined to give .

$$R_{Tf} = \frac{v_o}{i_i} = \frac{J_g}{i_i} \cdot \frac{v_o}{J_g} \cdot$$

$$= (\text{Current coupling factor})^{-1} \times R_{Tf}{}^*$$

$$= \left(\frac{r_g}{r_g + r_{if}}\right)^{-1} \cdot R_{Tf}{}^* \tag{5.47}$$

Numerically,

$$R_{Tf} = \left(\frac{2\ \text{k}\Omega}{2\ \text{k}\Omega + 22.4}\right)^{-1} \cdot (-9.66\ \text{k}\Omega) = -9.77\ \text{k}\Omega \tag{5.48}$$

The method described above for including the effects of r_g and r_L loading in the shunt-voltage feedback configuration is easily generalised to cover all four configurations. Experience of other analyses shows that symmetry exists between the methods for the four configurations without repeating each case. For series-input feedback (Fig. 4.1c and d) the input signal is a voltage and so a Thevenin equivalent circuit representation of the signal source is used, i.e. Z_g in series with an ideal voltage generator. In this case Z_g is incorporated in series with the input when forming the new amplifier circuit. Similarly, for current-output feedback (Fig. 4.1a and d) the load impedance Z_L is incorporated in series instead of in parallel with the output when forming the new amplifier circuit.

In general, the method for analysing a feedback amplifier circuit including the loading effects of the feedback block and the source and load impedances, consists of the steps outlined in Tables 5.1 and 5.2. Table 5.1 shows the steps required to construct and analyse the complete feedback circuit comprising the new forward-gain amplifier and used as an intermediate step in the calculation. Table 5.2 shows the subsequent steps required to calculate finally the closed-loop gain and the input and output impedances of the feedback circuit proper. The feedback amplifier circuit considered in the immediately previous Worked Example 5.5 is of the shunt voltage type and the reader should confirm that the second rows in these tables summarise the steps in that worked example. The other three rows in the tables give the corresponding steps to be carried out for similar analyses of the other three feedback configurations.

As with other tables for feedback amplifier circuits the entries in Tables 5.1 and 5.2 form a regular pattern. If any entry is covered up it can be correctly written down by considering other entries in the table. Try this yourself.

This completes the treatment of loading effects in feedback circuits. The development of the analysis of feedback circuits compared with the previous chapter has been that of achieving greater accuracy by making fewer assumptions about the circuit. Greater accuracy is obtained at the cost of lengthier calculations. In this chapter the assumption made about the feedback circuit is the reasonable one that both the forward amplifier and the feedback block are unilateral. What if these assumptions are not valid for a particular circuit? It should be realized that feedback amplifiers are electrical circuits, and like any circuit can be solved using standard circuit analysis techniques such as those based on mesh or nodal equations. From a designers' point of view recourse to such techniques may be necessary in difficult cases. However, they usually require greater calculation effort and also loss of insight into the circuit behaviour.

Table 5.1 Steps to be taken in Analysing the Hypothetical Closed-Loop Amplifier

Type of feedback	Ideal feedback to be calculated	Include block loading using	Form new forward-gain amplifier by combining	Analyse new amplifier to obtain the following open-loop quantities	Calculate the following closed-loop quantities of the hypothetical feedback amplifier		
					Gain	Input impedance	Output impedance
Shunt current	β_I	Fig. 5.3a	Z_g in parallel, Z_L in series	A_I^*, Z_i^*, Z_o^*	$A_{If}^* = \frac{A_I^*}{1 + A_I^*\beta_I}$	$Z_{if}^* = \frac{Z_i^*}{1 + A_I^*\beta_I}$	$Z_{of}^* = Z_o^*(1 + A_I^*\beta_I)$
Shunt voltage	β_G	Fig. 5.3b	Z_g in parallel, Z_L in parallel	R_T^*, Z_i^*, Z_o^*	$R_{Tf}^* = \frac{R_T^*}{1 + R_T^*\beta_G}$	$Z_{if}^* = \frac{Z_i^*}{1 + R_T^*\beta_G}$	$Z_{of}^* = \frac{Z_o^*}{1 + R_T^*\beta_G}$
Series voltage	β_V	Fig. 5.3c	Z_g in series, Z_L in parallel	A_V^*, Z_i^*, Z_o^*	$A_{Vf}^* = \frac{A_V^*}{1 + A_V^*\beta_V}$	$Z_{if}^* = Z_i^*(1 + A_V^*\beta_V)$	$Z_{of}^* = \frac{Z_o^*}{1 + A_V^*\beta_V}$
Series current	β_R	Fig. 5.3d	Z_g in series, Z_L in series	G_T^*, Z_i^*, Z_o^*	$G_{Tf}^* = \frac{G_T^*}{1 + G_T^*\beta_R}$	$Z_{if}^* = Z_i^*(1 + G_T^*\beta_R)$	$Z_{of}^* = Z_o^*(1 + G_T^*\beta_R)$

Table 5.2 Closed-Loop Expressions for the Feedback Amplifier Proper

Note: The expressions in this table use quantities obtained from Table 5.1.

Type of feedback	Closed-loop quantities of the feedback amplifier proper		
	Input impedance	Output impedance	Closed-loop gain
Shunt current	$Z_{if} = \frac{Z_g \times Z_{if}^*}{Z_g - Z_{if}^*}$	$Z_{of} = Z_{of}^* - Z_L$	$A_{If} = \frac{i_o}{i_i} = (\frac{Z_g}{Z_g + Z_{if}})^{-1} \cdot A_{If}^*$
Shunt voltage	$Z_{if} = \frac{Z_g \times Z_{if}^*}{Z_g - Z_{if}^*}$	$Z_{of} = \frac{Z_L * Z_{of}^*}{Z_L - Z_{of}^*}$	$R_{Tf} = \frac{v_o}{i_i} = (\frac{Z_g}{Z_g + Z_{if}})^{-1} \cdot R_{Tf}^*$
Series voltage	$Z_{if} = Z_{if}^* - Z_g$	$Z_{of} = \frac{Z_L * Z_{of}^*}{Z_L - Z_{of}^*}$	$A_{Vf} = \frac{v_o}{v_i} = (\frac{Z_{if}}{Z_g + Z_{if}})^{-1} \cdot A_{Vf}^*$
Series current	$Z_{if} = Z_{if}^* - Z_g$	$Z_{of} = Z_{of}^* - Z_L$	$G_{Tf} = \frac{i_o}{v_i} = (\frac{Z_{if}}{Z_g + Z_{if}})^{-1} \cdot G_{Tf}^*$

Frequency Response of Feedback Amplifiers

An important further beneficial effect of negative feedback is discussed in this section: the improvement of the frequency response of an amplifier. Consider the case of a forward amplifier having a mid-band gain A_{mb} and a single high-frequency effect which gives rise to an upper half-power point at ω_U. Thus

This is the standard form for a single high-frequency effect (see Equation 3.36).

$$A = \frac{A_{mb}}{1 + j\dfrac{\omega}{\omega_U}}$$

If feedback is applied to this amplifier using a feedback fraction β, then assuming β to be constant at all frequencies the closed-loop gain is given by

$$A_f = \frac{A}{1 + A\beta} = \frac{\dfrac{A_{mb}}{1 + j\dfrac{\omega}{\omega_U}}}{1 + \dfrac{A_{mb}}{1 + j\dfrac{\omega}{\omega_U}}}$$

Therefore

Multiplying numerator and denominator by the factor $(1 + j\dfrac{\omega}{\omega_U})$ and tidying up.

$$A_f = \frac{A_{mb}}{1 + j\dfrac{\omega}{\omega_U} + A_{mb}\cdot\beta} = \frac{A_{mb}}{(1 + A_{mb}\cdot\beta) + j\dfrac{\omega}{\omega_U}}$$

$$= \frac{A_{mb}}{(1 + A_{mb}\cdot\beta)}\cdot\frac{1}{1 + j\dfrac{\omega}{\omega_U(1 + A_{mb}\cdot\beta)}} \qquad (5.56)$$

This expression is now in the usual form of a constant mid-band gain term multiplied by a frequency-response term. From this the mid-band gain and upper half-power point after feedback has been applied can be recognized, and are:

$$A_{mbf} = \frac{A_{mb}}{1 + A_{mb}\cdot\beta} \qquad (5.57)$$

and

Once again the usual factor $(1 + A\beta)$ appears.

$$\omega_{Uf} = \omega_U(1 + A_{mb}\cdot\beta) \qquad (5.58)$$

The first of these two expressions shows that, as expected, the mid-band gain is divided by the factor $(1 + A_{mb}\cdot\beta)$ and is therefore reduced when the type of feedback is negative. The second expression shows that with negative feedback the upper half-power frequency is increased by the same factor. For large amounts of negative feedback the improvement can be considerable.

Exercise 5.3 Show that if negative feedback is applied to a forward amplifier having a single low-frequency effect with half-power point at ω_L, then

Hint: Use the same approach as that for ω_{Uf} in Equation 5.58.

$$\omega_{Lf} = \frac{\omega_L}{1 + A_{mb}\cdot\beta} \qquad (5.59)$$

The effect of negative feedback is to improve both the lower and upper half-power frequencies. The improvement becomes greater as the amount of negative

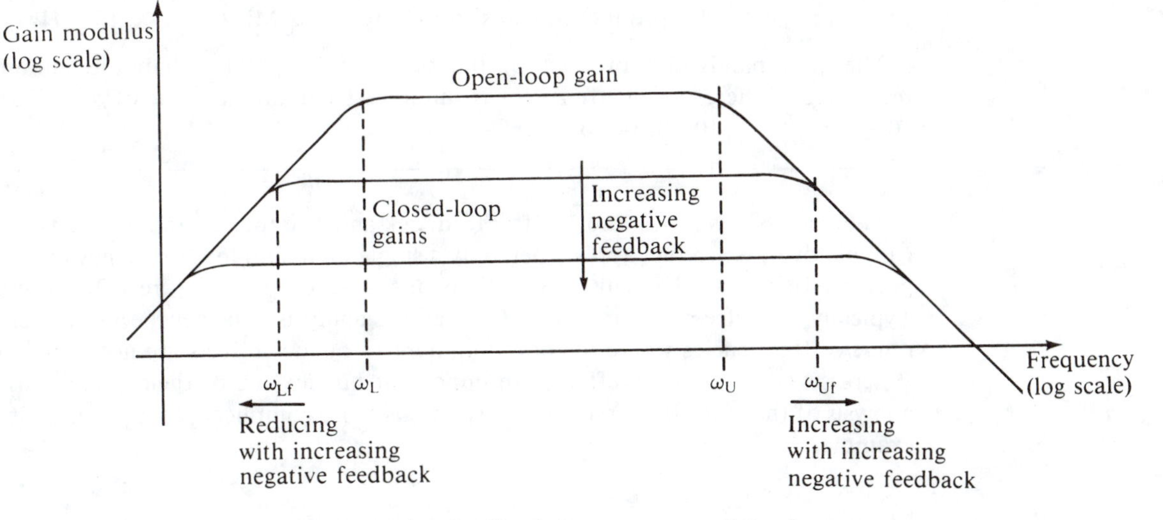

Fig. 5.10 Negative feedback improves amplifier frequency response.

feedback is increased. This property is illustrated in Fig. 5.10. Note as $(1 + A_{mb} \cdot \beta)$ is increased the value of the closed-loop gain is decreased and the bandwidth is widened.

This can also be understood in terms of reduced sensitivity provided by negative feedback. In the frequency range ω_{fL} to ω_{fU} the open-loop gain A varies with frequency. Negative feedback reduces the effects of those variations and tends to hold the closed loop gain constant. At the limits ω_{fL} and ω_{fU} the feedback is too weak to maintain this and the closed-loop gain approximately equals the open-loop gain.

A useful figure-of-merit is the *gain-bandwidth product*, defined as

$$\text{Gain-bandwidth product} = \text{Mid-band gain} \times \text{Bandwidth}$$

$$= \text{Mid-band gain} \times (\omega_U - \omega_L)$$

(Note that *gain* here means ratios between currents and/or voltages and *not* power ratios.) In most practical cases $\omega_U >> \omega_L$ and so

$$\text{Gain-bandwidth product} \simeq \text{Mid-band gain} \times \omega_U$$

Substituting for mid-band gain and ω_U after feedback using Equations 5.57 and 5.58 then

$$\text{Gain-bandwidth product} = \frac{A_{mb}}{(1 + A_{mb} \cdot \beta)} \cdot \omega_U (1 + A_{mb} \cdot \beta)$$

$$= A_{mb} \cdot \omega_U = \text{constant} \quad (5.60)$$

The gain-bandwidth is therefore a constant which is defined by the parameters of the amplifier before feedback is applied and is independent of the fraction β. It is useful as a figure of merit because it can be used to indicate if a specification can be met by a given forward amplifier, as illustrated by the following worked example.

Or, conversely, what the parameters of the forward amplifier must be to achieve a desired closed-loop performance.

Worked Example 5.6

A transistor has a current gain of 100 and an upper half-power frequency of 2 MHz. A closed-loop current gain of 10 is required with a bandwidth of 10 MHz or better. Is the transistor suitable and what value of β is required?

Solution:

$$\text{Gain bandwidth product required} = 10 \times 10\ \text{MHz} = 10^8 \cdot \text{Hz}$$

$$\text{Gain bandwidth product of transistor} = 100 \times 2\ \text{MHz} = 2 \times 10^8 \cdot \text{Hz}$$

The gain bandwidth product of the transistor is greater than the minimum required and therefore the transistor is suitable. The required value of β is given by $10 = 100/(1 + 100\beta)$, i.e. $\beta = 0.09$.

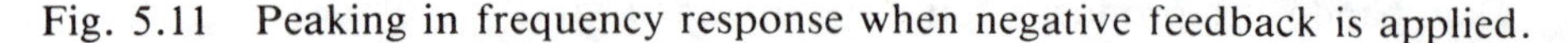

The expressions for ω_L and ω_U after feedback and also the constant property of the gain bandwidth product apply when only a single high- and low-frequency effect are present. In practical circuits more than one frequency effect are often present. Typically one effect dominates and the above analysis can then be used over the range of frequencies where the non-dominant frequency effects are not significant. Where several frequency effects are important but no one of them dominates the analysis of the amplifier, frequency response is more complicated than in this simple example.

Instability

When instability was discussed in Chapter 2 it arose because of positive feedback. When the forward amplifier feedback is intended to be negative phase-shifts can occur which cause the feedback to become positive in some parts of the amplifier's frequency range.

Consider what happens when negative feedback is applied to a forward amplifier with three stray-capacitance high-frequency effects. When feedback is applied the resulting closed-loop frequency response is typically of the form shown in Fig. 5.11. The mid-band gain is reduced and the general bandwidth is increased as expected for negative feedback. However, strong peaking of the response has appeared in the region of the half-power frequency. Generally a flat frequency response is preferred and so this peaking is undesirable. The presence of the peak takes on greater significance when it is realized that for signal frequencies in the region of the peak the gain with feedback is greater than the gain without feedback. Therefore, in this

Peaking can also occur in a similar way at the low-frequency end of the amplifier response.

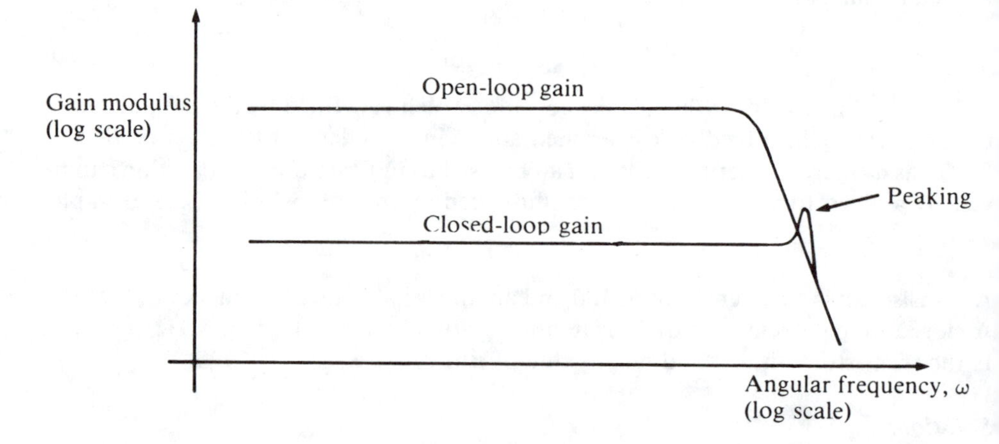

Fig. 5.11 Peaking in frequency response when negative feedback is applied.

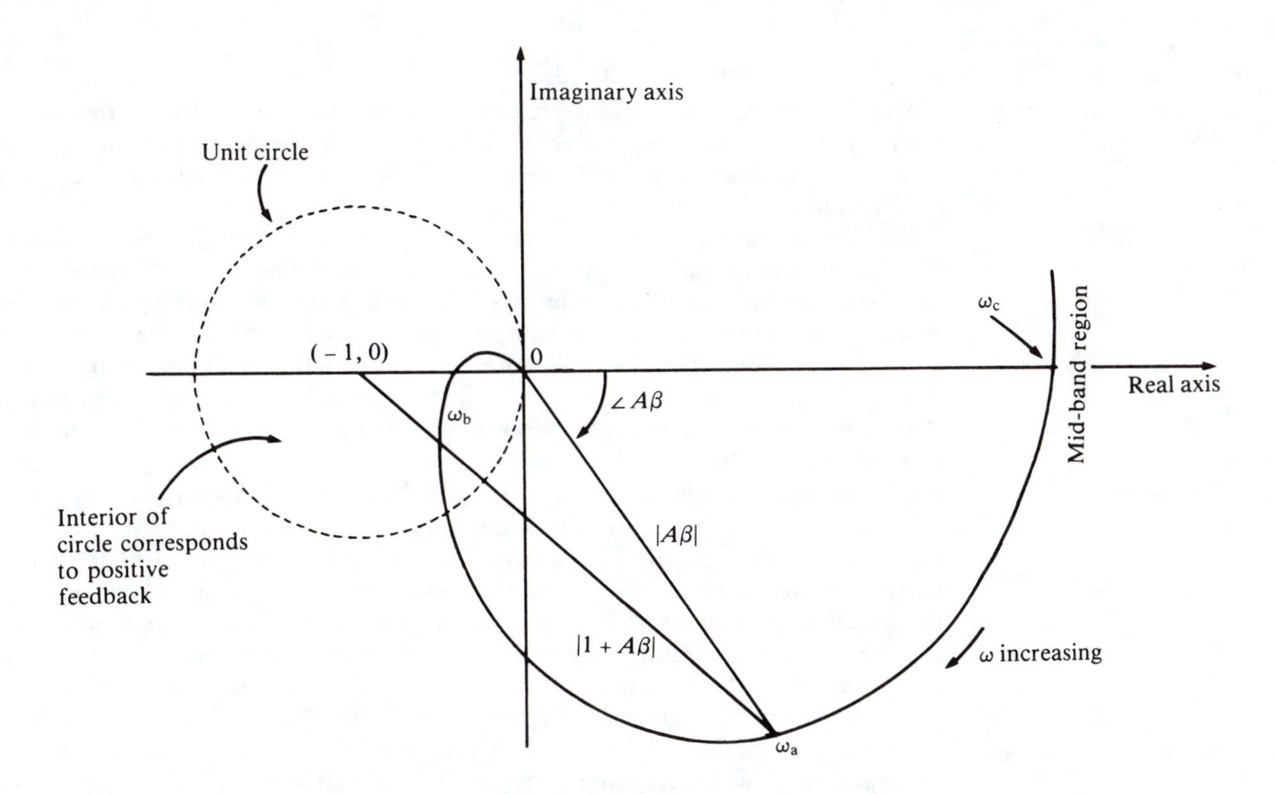

Fig. 5.12 Nyquist plot of A_β for amplifier with three high-frequency effects.

region the feedback must be positive and consequently all the disadvantages of positive feedback such as increased sensitivity and distortion are to be found. The cause of this positive feedback is analysed below.

The reason for the occurrence of the peak in the frequency response is found by examining the Nyquist plot of $A\beta$, i.e. the plot of $A\beta$ in the complex plane, for all signal frequencies. The forward gain A for the case of three equal high-frequency effects was shown in Equation 3.38 and plotted in Fig. 3.8b. For the feedback arrangements discussed so far, the feedback block is constructed from one or two passive resistors and so β is assumed to be a constant which depends on the resistance values but not on signal frequency. Consequently the plot of $A\beta$ is the same general shape as for A in Fig. 3.8b but is scaled in size by the constant β. The Nyquist plot of $A\beta$ is typically as shown in Fig. 5.12. The figure concentrates on the higher-frequency portion of the locus where the peaking behaviour occurs.

The number of resistors depends on the circuit configuration being used, see Figs 4.4 to 4.7. The frequency-response behaviour discussed here applies generally to all configurations.

The locus is of $A\beta$ and for a given angular frequency point, say ω_a the distance from the origin to the point, $0\omega_a$ is equal to the modulus of $A\omega$. The phase angle $\underline{/A\omega}$, is the angle which 0ω moves through, starting from the horizontal axis at frequencies in the mid-band region as the frequency is increased to the point ω_a. As the frequency is increased, the effect of the capacitances in the circuit causes the phase lag $\underline{/A\beta}$ to increase while the modulus $|A\beta|$ reduces, thus the locus moves clockwise and in the limit at infinite frequency reaches the origin 0.

A single high-frequency effect produces a limiting phase shift of −90° at infinite frequency so three effects give a limiting phase shift of −270°. The locus therefore moves clockwise through three quadrants before disappearing at the origin.

The type of feedback is determined by the magnitude of the factor $(1 + A\beta)$. The magnitude of the factor $(1 + A\beta)$ can be interpreted as the distance between any point on the $A\beta$ locus and the point $1\underline{/180^\circ}$, as shown in the figure, because

Note $1\underline{/180^\circ} \equiv -1 + j0$.

$$|1 + A\beta| = |A\beta - (-1)| = \left| |A\beta| \underline{/A\beta} - 1 \underline{/180°} \right| \quad (5.61)$$

The unit circle is drawn with a broken line in Fig. 5.12.

All points inside a unity radius circle drawn round the point (– 1,0) correspond to values of $|1 + A\beta|$ being less than unity. Referring to the basic definitions in Chapter 2 this means that the feedback at any such point inside the unit circle is positive and not negative.

The reason for the peak in the frequency response in Fig. 5.11 is now clear. Considering a point such as ω_b the radius drawn from the point (– 1,0) to the point ω_b is less than unity. Hence the feedback is positive and the resulting closed-loop gain is greater than the amplifier gain without feedback at this angular frequency. As the angular frequency is increased it passes through the region including ω_b, and $|1 + A\beta|$ reduces, reaches a minimum and then increases again. This gives rise to a peak in the closed loop response because the closed loop gain is equal to the open loop gain divided by $|1 + A\beta|$.

General behaviour.

This behaviour applies not only to an amplifier with three high-frequency effects as mentioned above but also to any amplifier in which a portion of the loop gain locus falls inside the unity radius circle. An amplifier with a single high- or low-frequency effect cannot show the peaking effect with feedback applied (assuming β to be a constant without phase angle) because the greatest phase deviation is $\pm 90°$ and the $A\beta$ locus is confined to the right half of vertical axis on the Nyquist plot (see Fig. 3.8a). However, the presence of two or more high- or low-frequency effects in an amplifier results in phase shifts approaching $\pm 180°$, or more. In this case $A\beta$ can enter the unit circle and over certain ranges of frequency positive feedback is present. Often this is acceptable provided the $A\beta$ locus does not penetrate the unit circle any great distance, thus keeping any peaking in the frequency response to an acceptable level.

In practice a small amount of positive feedback is sometimes deliberately used to lift the closed-loop gain a little in the half-power region and thereby extend the amplifier bandwidth even further.

The amplifier behaviour discussed so far can result in undesirable peaking of the frequency response but the amplifier is not unstable. An amplifier which is unstable contains signals which circulate round the feedback loop without dying away even when the input signal is removed. This means that the circuit oscillates and actually *generates* an alternating waveform. Considering Fig. 5.12 again, if the feedback fraction β is increased the locus $A\beta$ grows in size radially from the origin 0. Points in the region of ω_b move towards the point (– 1,0), thus reducing $|1 + A\beta|$ and increasing the peaking in the closed-loop response. If β is increased so that the locus passes exactly through the point (– 1,0) then at that point $|1 + A\beta|$ is zero and the closed loop gain is thus infinite. At this point the loop gain is unity, and the signals in the loop are self-sustaining without needing an input signal. The output signals usually are oscillatory in nature with an angular frequency where the A locus crosses the (– 1,0) point. This marks the onset of amplifier instability. As β is increased further so that the locus encloses the (– 1,0) point, oscillations build up in amplitude until limited by non-linearities in the amplifier circuitry.

To appreciate the connection between the Nyquist plot of Fig. 5.12 and the discussion in Chapter 2 summarised in Fig. 2.3, consider the operation at the mid-band angular frequency ω_c where the forward amplifier phase shift is zero. At this point the product $A\beta$ is a positive number. In order to allow a comparison with Fig. 2.3 assume that A is positive and fixed in value, and consider the effect of varying the magnitude and sign of β. The case actually drawn in Fig. 5.12 (at the point ω_c) corresponds to some point in the negative feedback region (i) in Fig. 2.3. If the positive value of β is reduced towards zero the ω_c point in Fig. 5.12 moves along the axis to the origin. In Fig. 2.3 this corresponds to movement along region (i) to the boundary with region (ii). If β is now increased from zero in the negative direction the ω_c point moves into the unit circle and gives positive feedback as in region (ii) of Fig. 2.3. When β has the value which makes $A\beta = -1$, thus placing the ω_c point at the point $(-1,0)$, this gives rise to infinite closed-loop gain as in region (iii). As β increases further in this direction the ω_c point approaches the unit circle boundary where the system passes from region (iv) to region (v) of Fig. 2.3 and the feedback changes back from positive to negative.

Worked Example 5.7

Negative feedback is applied to a three-stage amplifier having an overall mid-band gain A_{mb}, each stage of which has an equal half-power angular frequency of ω_1. Derive expressions for the value of the feedback fraction β (assumed to have zero phase shift) which just causes instability and also for the expected frequency of oscillation.

Solution: The forward gain is of the form

$$A = \frac{A_{mb}}{(1 + j\dfrac{\omega}{\omega_1})^3}$$

(see Equation 3.38). At the limiting value of β, for some angular frequency ω_o, the locus of $A\beta$ exactly passes through the point $(-1,0)$, i.e. $-1 + j0$ in complex notation. When this occurs

$$A\beta \equiv -1 + j0$$

Substituting for A

$$\frac{A_{mb}\cdot\beta}{(1 + j\dfrac{\omega_o}{\omega_1})^3} \equiv -1 + j0$$

Multiplying both sides by the cubed term and expanding,

$$A_{mb} \equiv (-1 + j0)\,(1 + j\frac{\omega_o}{\omega_1})^3 \equiv -1 - j3\frac{\omega_o}{\omega_1} + 3\frac{\omega_o^2}{\omega_1^2} + j\frac{\omega_o^3}{\omega_1^3}$$

Separating real and imaginary terms

$$(1 + A_{mb}\cdot\beta - 3\frac{\omega_o^2}{\omega_1^2}) + j(3\frac{\omega_o}{\omega_1} - \frac{\omega_o^3}{\omega_1^3}) \equiv 0$$

To satisfy this condition both real and imaginary parts have to be simultaneously equal to zero. Taking first the imaginary part,

Two solutions for ω_0. We examine each one.

$$3\frac{\omega_o}{\omega_1} - \frac{\omega_o^3}{\omega_1^3} = 0 \tag{5.62}$$

Thus $\omega_o = 0$ or $\omega_o = \sqrt{3}\,\omega_1$. Thus two frequencies satisfy this condition. Consider now the real part

$$1 + A_{mb}\cdot\beta - 3\frac{\omega_o^2}{\omega_1^2} = 0 \tag{5.63}$$

Taking first the condition $\omega_o = 0$, in Equation 5.62 then Equation 5.63 becomes

$$1 + A_{mb} \cdot \beta = 0$$

thus $A_{mb} \cdot \beta = -1$. This solution can be discounted because at mid-band frequencies for the feedback to be negative, as assumed, the product $A_{mb} \cdot \beta$ must be positive.

Taking the other solution for ω_o in Equation 5.62 and substituting in Equation 5.63,

An amplifier with two high-frequency (or low-frequency) effects cannot oscillate with negative feedback. To oscillate its $A\beta$ locus must intersect $(-1,0)$ which requires a phase shift of 180°

$$1 + A_{mb} \cdot \beta - 3(\sqrt{3})^2 = 0$$

from which $A_{mb} \cdot \beta = +8$. This is positive and can be accepted. It can be concluded from this that the limiting value is $\beta = 8/A_{mb}$. To avoid instability β has to be less than this. If the limit is reached then sustained oscillations occur at a frequency given by $\omega_o = \sqrt{3}.\omega_1$.

Now try the following exercise.

Exercise 5.4 Confirm that at the angular frequency of oscillation ω_o in the previous worked example the phase shift lag in each stage of the amplifier is 60°. Do this by substituting the value of ω_o into the gain expression for each stage of the amplifier.

The previous worked example indicates that whether or not the three-stage amplifier (having the locus shown in Fig. 5.12) is stable depends on β. If β is small the amplifier is stable. If β is increased the locus expands until it touches the $(-1,0)$ point and the amplifier is then unstable. Because the system stability depends on β it is said to be *conditionally stable*. Not every $A\beta$ locus is conditionally stable. An example of an unconditionally stable locus is shown in Fig. 5.13a, where the locus nowhere crosses the negative real axis and cannot therefore be made to touch the $(-1,0)$ point by increasing β.

It might be thought that if the $A\beta$ locus crosses the negative real axis to the left of the $(-1,0)$ point as, for example, in Fig. 5.13b and c, the system is bound to be unstable. However, it happens that the system represented by the locus of Fig. 5.13b is unstable, while that represented by Fig. 5.13c is stable. This follows because for instability to occur the locus must *encircle* the point $(-1,0)$, which is what happens in Fig. 5.13b but not in Fig. 5.13c.

This stability criterion is a classical result due to Nyquist, the proof being too long to give here.

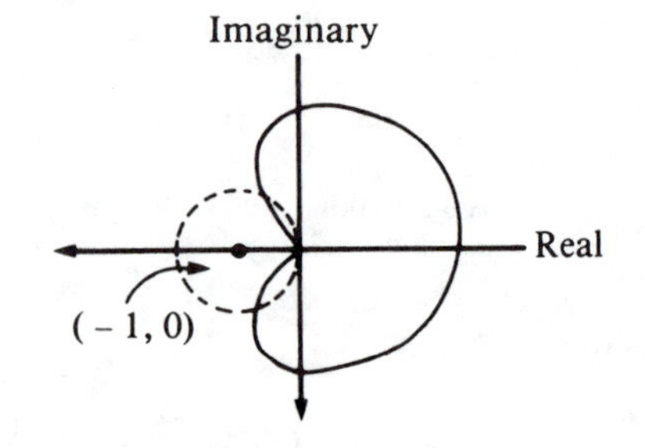

(a) Unconditionally stable

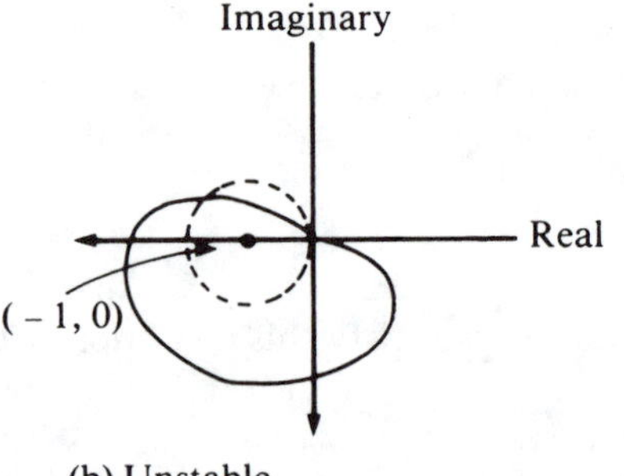

(b) Unstable

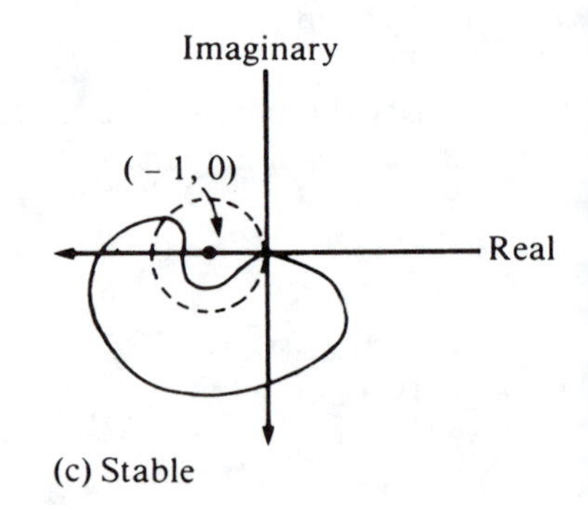

(c) Stable

Fig. 5.13 More Nyquist diagrams.

Conditionally stable amplifiers with $A\beta$ loci which cross the negative real axis to the left of the $(-1,0)$ point but which nevertheless are stable, as in Fig. 5.13c, are mainly of theoretical interest because, in practice, difficulties can arise and therefore they are not normally used. One difficulty occurs if the locus is designed to pass close to the $(-1,0)$ point. In this case manufacturing variations in components could cause the loci of some individual amplifier specimens to deviate from the design value and pass on the wrong side of the $(-1,0)$ point thus enclosing it and causing instability. A second difficulty can occur when the power supplies to an amplifier are switched on. As the circuit biasing conditions settle down the gain of the forward amplifier could increase progressively through a low value to its normal value. In this case the $A\beta$ locus of the circuit begins at a region close to the origin in the Nyquist diagram and expands out towards its normal value. In doing so, one of the loops in the locus encloses the $(-1,0)$ point. Oscillations may then take over and prevent the circuit reaching a stable amplifying condition.

In modern feedback circuit practice, because amplifying devices are cheap there is a tendency to design circuits with large feedback factors $(1 + A\beta)$ to obtain the greater benefits that result. The $A\beta$ loci on the Nyquist plots can be quite large with increased risk of encircling the $(-1,0)$ point. So without some modification many modern negative feedback circuits would be unstable. The remedy is to apply *frequency compensation*. Here the frequency responses of the forward amplifier and/or the feedback block are modified by adding capacitors or other frequency-dependent components to shape the $A\beta$ locus so that it does not enclose the critical $(-1,0)$ point.

Frequency compensation is dealt with, for example, in Millman, J. *Microelectronics* (McGraw-Hill, 1979), also, Roberge, J.K. *Operational Amplifiers: Theory and Practice* (Wiley, 1975). Frequency compensation plays an important part in automatic control, see for example, Healey, M. *Principles of Automatic Control* 3rd ed., (EUP, 1975).

Summary

In practical circuits the feedback block may load the forward amplifier because it draws a finite current, for voltage-output sampling, or develops a finite voltage drop for current-output sampling. Similarly if the input impedance of an amplifier is not very low for shunt feedback, or not very high for series feedback, it loads the output of the feedback block. Both of these may affect the performance of the feedback circuit significantly. A convenient way to analyse a circuit, taking into account these effects, is to absorb the causes of loading into the forward amplifier as shown in Fig. 5.3. The methods of analysis previously developed in Chapter 4 and based on Tables 4.1 and 4.2 then apply.

The presence of the signal source impedance Z_g at the input and the load impedance Z_L at the output of the feedback circuit can also affect the performance of the feedback amplifier. Again for the purposes of analysis these causes of loading can be absorbed into the forward Z_g and Z_L are removed from the closed-loop input and output impedances to obtain closed-loop input and output impedance Z_{if} and Z_{of}, for the actual feedback circuit. The closed-loop gain is multiplied by the inverse of a coupling factor to obtain the closed-loop gain of the actual feedback circuit. These procedures for all feedback configurations are summarised in Tables 5.1 and 5.2.

A separate subject discussed is the effect of feedback on amplifier bandwidth. For a single high-frequency effect the bandwidth is increased by the factor $(1 + A\beta)$. Mid-band gain multiplied by bandwidth is independent of β and this, the *gain-bandwidth* product, is a useful figure of merit.

The application of negative feedback does not always produce beneficial effects on the frequency response. Peaking in the response can occur. Examination of the Nyquist plots shows that this occurs because over certain ranges of signal frequency the feedback in fact is positive. This happens when the locus of $A\beta$ enters the unit circle centred on $(-1,0)$.

If the $A\beta$ locus touches, or encircles, the $(-1,0)$ point, i.e. $A\beta \equiv -1 + j0$, sustained oscillations occur. If the $A\beta$ locus goes behind the $(-1,0)$ point but does not encircle it the amplifier is theoretically stable. However, for practical reasons amplifiers of this type are generally not used.

Problems

Neglect Z_g and Z_L loading effects in Problems 5.1–5.4.

5.1 Repeat Worked Example 5.1 using a value of $R = 1$ kΩ.

5.2 Re-analyse the series-voltage feedback circuit of Worked Example 5.2. This time better transistors are used having $h_{fe} = 200$. Other parameters are unchanged.

5.3 Series-current feedback is applied to an amplifier using the arrangement in Fig. 4.6a. For the forward amplifier $r_i' = 1$ kΩ, $r_o' = 1$ kΩ, and the transconductance parameter is -1 S. The feedback resistor is $R = 1$ kΩ. Calculate the closed-loop input and output impedances and the closed-loop gain. Take account of loading between the forward amplifier and the feedback block.

5.4 Repeat the last problem for the circuit in Fig. 4.7a given that $r_i' = 10$ kΩ $r_o' = 10$ kΩ, $K_I = 10^5$, $R_1 = 10$ kΩ and $R_2 = 100$ Ω.

5.5 Repeat Problem 5.3 taking into account Z_g and Z_L loading effects. The source and load resistances are 1 kΩ and 250 Ω.

5.6 Series-voltage feedback is applied to an amplifier as shown in Fig. 4.5a. The amplifier parameters are $r_i' = 10$ kΩ, $r_o' = 10$ kΩ and $K_V = 10^5$. The feedback resistors are $R_1 = 2$ kΩ, $R_2 = 400$ Ω. The source and load resistances are $r_g = 50$ kΩ, $r_L = 1$ kΩ. Calculate the closed-loop input and output impedances and also the closed-loop gain. Take account of all loading effects.

5.7 An amplifier has a mid-band gain of 40 dB and upper and lower half-power frequencies (each owing to a single effect) at 100 Hz and 20 kHz. Negative feedback is applied to lower the mid-band gain to 20 dB. Calculate the new upper and lower half-power frequencies.

5.8 Repeat Worked Example 5.7, assuming that the amplifier has three *low*-frequency effects each having a cutoff frequency of ω_L.

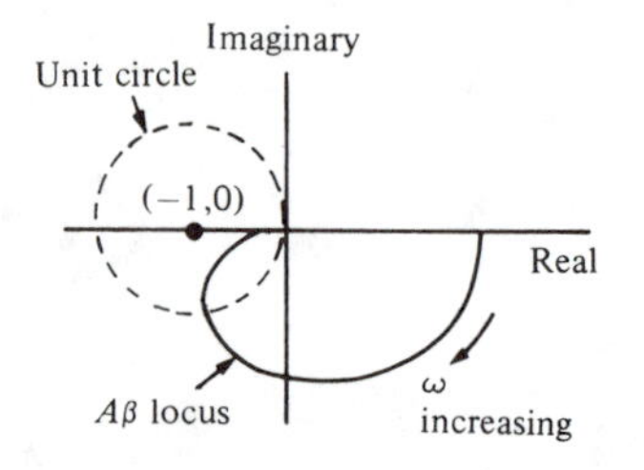

5.9 An amplifier has a mid-band gain A_{mb} and two high-frequency effects each having a corner frequency of ω_1. Negative feedback is applied having a feedback fraction β which is constant with frequency. Thus the Nyquist diagram of the $A\beta$ locus is shaped as shown in the margin. Derive an expression for the signal frequency beyond which the feedback becomes positive.

5.10 Repeat Problem 5.9 when the two effects are of low-frequency type.

The Op. Amp. – Basic Ideas and Circuits

6

Objectives

- ☐ To describe the distinguishing features of an op. amp.
- ☐ To explain how the inverting voltage amplified op. amp. circuit works.
- ☐ To appreciate that the performance of negative feedback op. amp. circuits is almost entirely defined by the components surrounding the op. amp. and is negligibly affected by the gain and input and output impedances of the op. amp.
- ☐ To appreciate the versatility of the inverting voltage amplifier circuit by seeing some useful circuits derived from it, including integrators.
- ☐ To describe a second basic op. amp. circuit, the non-inverting voltage amplifier and to know of its important impedance buffering property.
- ☐ To choose and connect power supplies to the op. amp. to make a working circuit.

What is an Operational Amplifier?

An operational amplifier, or op. amp., is an amplifier which responds only to the difference between two input voltages and has a very large voltage gain, and high input resistance.

Operational amplifiers were first conceived for use in analogue computers. In an analogue computer, op. amps. and other components are connected to make circuit modules which quite precisely perform mathematical operations on electrical signals which are the analogues of physical quantities. (Hence the name *operational* amplifier.) These operations include addition, scaling and integration. The circuit modules (or building blocks) are interconnected in specific ways to simulate some physical problem such as the behaviour of a vehicle suspension. It was recognized early on that op. amp. circuits could be useful not only for analogue computers but for analogue signal processing in general. However, the relatively high cost of the op. amps. using thermionic valves was an obstacle to their widespread use. The situation changed dramatically in the 1960s with the ready availability of cheap integrated circuits on silicon chips.

One of the most popular op. amps. is the 741-type, which has been produced by several manufacturers for many years. Also several improved versions have been developed having different internal circuitry to the original 741 op. amps. but which can directly replace it. Such devices are said to be directly *pin-compatible*. The 741 is plastic encapsulated with *dual-in-line* (DIL) rows of pins, as shown in Fig. 6.1a. The pin connections are shown in Fig. 6.1b. The commonly used triangular symbol for the differential amplifier is shown and the inverting and non-inverting inputs to the op. amp. can be seen to be at pins 3 and 2 respectively. As explained in Chapter 3, if a voltage waveform is applied to pin 2 it is inverted at the output, but if applied to

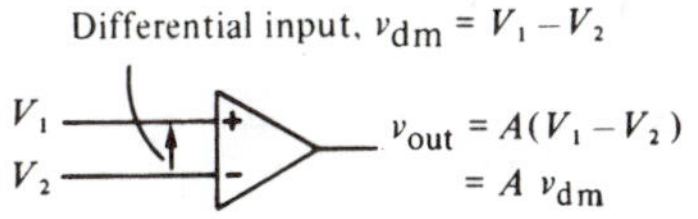

Differential input, $v_{dm} = V_1 - V_2$

V_1 +, V_2 −

$v_{out} = A(V_1 - V_2)$
$= A\, v_{dm}$

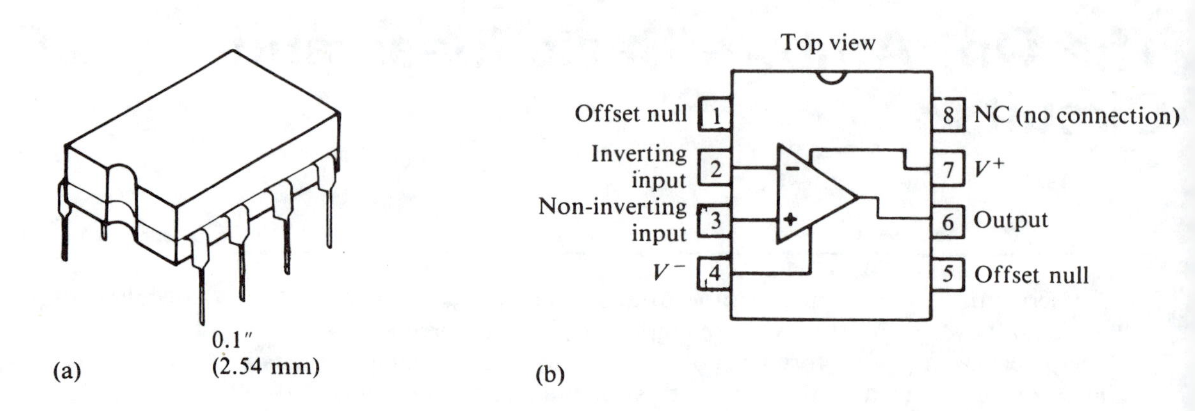

Fig. 6.1 Integrated circuit 741. Operational amplifier: (a) DIL package, (b) pin connections.

pin 3 it does not (in each case the other terminal would be earthed). The op. amp. output appears at pin 6. Several other pins are labelled in the figure. The most important of these are 7 and 4, where positive and negative constant supply voltages, V^+ and V^- are connected to supply energy to the integrated circuit. This power supply enables the amplifier to provide greater output signal power than is absorbed at its signal input terminals. Without these connections the amplifier does not work. Two other pins are provided on the 741, labelled *offset null*. Offset effects are one of several types of non-ideal behaviour in the op. amp. and are dealt with in the next chapter.

Other device packages are encountered but the DIL is probably the most common. Also some IC packages contain more than one op. amp. (for example the 747 type containing two 741 op. amps.).

Many types of op. amps. are manufactured. Some have advantages over others such as better electrical characteristics wider operating temperature range, or less cost. All op. amps. are similar in that they are differential amplifiers having high differential-mode gain. The internal circuitry to realize an operational amplifier also varies. To use operational amplifiers successfully it is not important to understand the internal circuit operation. However, the general form of an op. amp. internal circuitry is in three main sections as shown in Fig. 6.2. The first is a differential amplifier circuit, an example is as shown in Fig. 4.5b, which is followed by one or more stages of transistor amplification. The function of the final section is primarily to provide the output voltage signal and sufficient current to drive the external load resistance. Table 6.1 also shows typical values for the 741-type op. amp. and for another type, the CA3130E op. amp.

Using the differential amplifier notation introduced in Chapter 3, the voltage gain A is really the differential mode amplification A_{dm}.

Table 6.1

	Ideal op. amp.	741	CA3130E
Voltage gain, A	∞	200 000	300 000
Input resistance	∞ Ω	2 MΩ	1.5×10^{12} Ω
Output resistance	0 Ω	75 Ω	—

The CA3130E has a significantly higher input resistance than the 741-type because the differential input stage inside the CA3130E uses field-effect transistors which have significantly lower input currents than the bipolar transistors in the 741 op. amp. This feature is advantageous in some op. amp. applications.

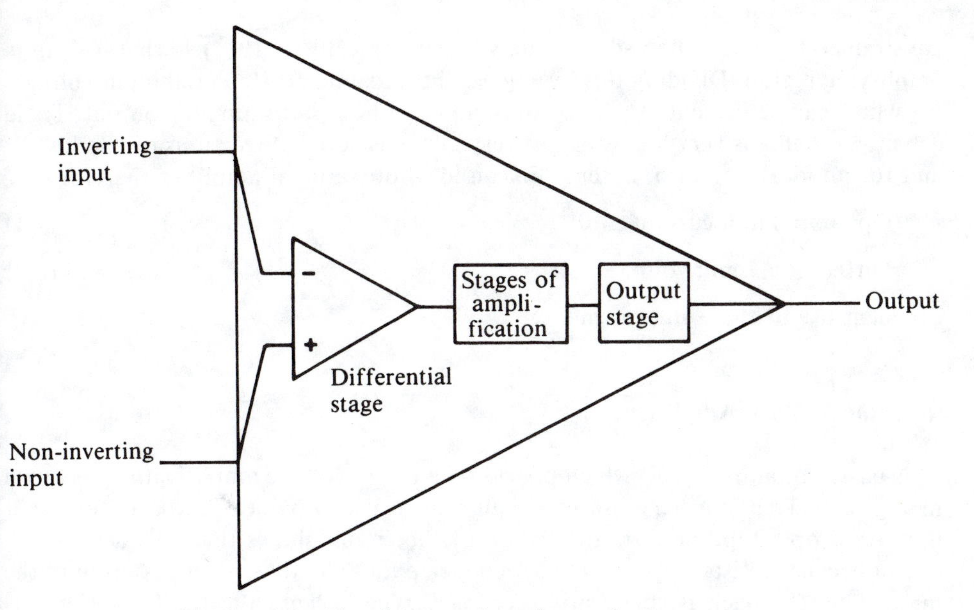

Fig. 6.2 Usual internal sections of an op. amp.

The principal point to notice is the very high value of the gain A for the practical op. amps. This leads to high performance negative feedback circuits.

In the following chapters the effects of op. amp. non-idealities are discussed and a wider range of more advanced op. amp. circuits is described. Non-idealities are important but they generally contribute second-order effects to circuit performance. The performance of a circuit can usually be calculated quite accurately by assuming the op. amp. to be ideal. In this case, circuit analysis is aided by using two simple rules. The first of these follows from the property of infinite input impedance of an ideal op. amp. Thus the current drawn by the op. amp. input terminals is assumed to be zero. The second follows from the consideration of the signal voltages (see Fig. 6.3). In a practical voltage amplifier the output voltage of a op. amp. is

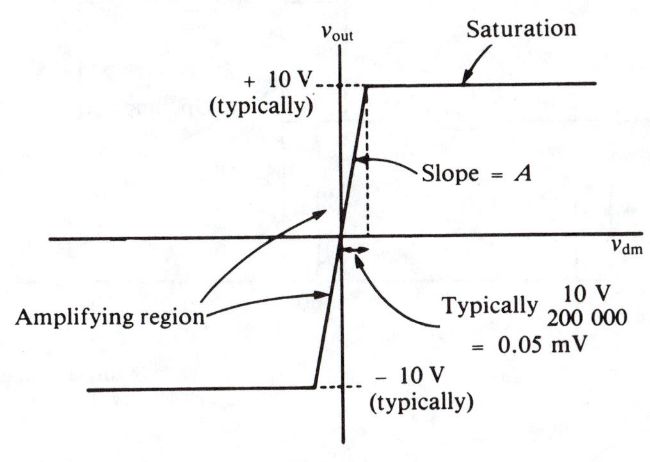

Fig. 6.3

constrained to be less than some finite voltage (typically $\pm$ 10 V) so that it is in its amplifying region. Dividing this by A gives the maximum differential input voltage, v_{dm} which can be present at the op. amp. input without saturating the output. Table 6.1 shows that A is very high so v_{dm} is very low (measured in fractions of a millivolt) and for an ideal op. amp. is zero. For an ideal operational amplifier therefore

$$\text{Op. amp. input current} = 0 \tag{6.1}$$

$$\text{Differential input voltage} = 0 \tag{6.2}$$

Frequent use of these rules is made.

Inverting Voltage Amplifier

As is common practice this diagram does not show the other op. amp. connections, such as power supplies V^+ and V^-, but which nevertheless are required for a practical working circuit.

This applies to most op. amp. circuits. The op. amp. is just a part of the whole circuit.

This basic op. amp. circuit which provides inverting voltage amplification is shown in Fig. 6.4. This is a very useful circuit which also provides the basis for other important op. amp. circuits. Inspection of the circuit shows that it is a feedback circuit because resistor R_2 provides a feedback path from the op. amp. output to the input. The feedback is of negative feedback type because the feedback signal is connected to the inverting terminal (labelled '$-$'). The non-inverting terminal (labelled '$+$') is connected to the common rail running through the circuit. A significant point to remember is that the op. amp. input terminals are not the place where the main input signal is applied. The main input signal is applied between R_1 and the common rail.

The operation of this circuit can be analysed and understood without using feedback theory. Assume an ideal op. amp. and commence by writing the equations for the circuit which surrounds the op. amp. At node X the currents satisfy Kirchhoff's current law. Thus

$$i_1 = i_2 + i_a \tag{6.3}$$

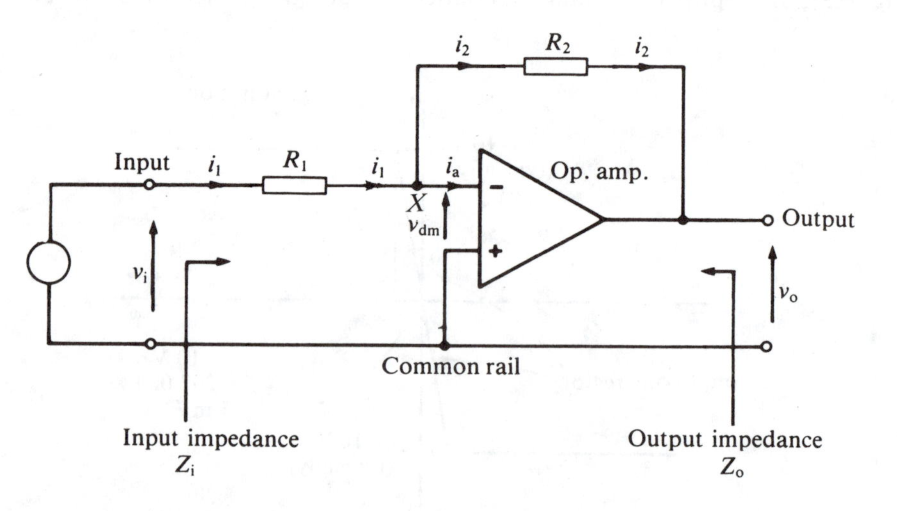

Fig. 6.4 Inverting voltage amplifier.

Using Ohm's law with resistances R_1 and R_2, this equation is rewritten,

$$\frac{v_i - v_{dm}}{R_1} = \frac{v_{dm} - v_o}{R_2} + i_a \qquad (6.4)$$

When using Ohm's law like this be careful to satisfy the sign convention. The voltage difference is taken between the voltage where the current arrow goes into the resistor (in general impedance), minus the voltage where it comes out.

Next account for the effect of the op. amp. Because it is assumed to be ideal, Rules 6.1 and 6.2 are applicable and so $v_{dm} = 0$ and $i_a = 0$. Because in practice $v_{dm} \approx 0$, the inverting input terminal is very nearly at common rail potential and is called the *virtual earth* point.

Thus Equation 6.4 becomes

$$\frac{v_i}{R_1} = i_1 = i_2 = -\frac{v_o}{R_2}$$

The basic reason for this important simple relation between v_i and v_o is that the two currents flowing into the virtual earth point are equal and opposite. A simple rearrangement gives the expression for the overall voltage gain of the circuit,

$$\text{Overall voltage gain, } A_v = \frac{v_o}{v_i} = -\frac{R_2}{R_1} \qquad (6.5)$$

The negative sign in this expression mean that a waveform $v_i(t)$ applied at the input to the circuit appears inverted at the output. The amplitude of the signal can be scaled by any desired amount by choosing values of R_1 and R_2 to give the desired A_v according to Equation 6.5.

It should be remembered that there is a simple relationship between v_o and i_1; $v_o = -R_2 i_1$. Thus the part of the circuit shown in Fig. 6.4 comprising only R_2 and the op. amp. can be used as a current-voltage convertor. This is the trans-resistance amplifier, one of the four basic types of amplifier presented in Chapter 3.

The reader may also recognise this part of the circuit as an example of shunt-voltage feedback as defined in Chapter 4.

The input and output impedances of the inverting amplifier (Fig. 6.4) circuit are also of interest. The output impedance of an ideal op. amp. is zero and therefore the output impedance of the full circuit is zero.

Chapter 4 shows that because the feedback here is voltage output type the small but finite output impedance of a practical op. amp. is very much reduced by feedback.

Output impedance, $Z_o = 0\ \Omega$

The input impedance of the circuit is defined by the ratio $Z_i = v_i/i_i$. But

$$i_1 = \frac{v_1 - v_{dm}}{R_1}$$

and for an ideal op. amp. $v_{dm} = 0$. Thus

$$i_1 = \frac{v_1}{R_1}$$

and so input impedance, $Z_i = R_1$.

Worked Example 6.1

Design an op. amp. inverting voltage amplifier to have an input resistance of 1 kΩ and a gain of -100.

Solution: The design requires the calculation of suitable values of R_1 and R_2 in Fig. 6.4. Directly from Equation 6.7

$R_1 = Z_i = 1\ \text{k}\Omega$

The gain expression (Equation 6.5) gives

$$\text{Voltage gain, } A_v = -\frac{R_2}{R_1}$$

and substituting values, gives

$$-100 = -\frac{R_2}{1\ \text{k}\Omega}$$

thus $R_2 = 100\ \text{k}\Omega$. Hence, the design values are $R_1 = 1\ \text{k}\Omega$, $R_2 = 100\ \text{k}\Omega$.

Bias currents and other non-idealities which limit the usable range of resistances are dealt with in the next chapter.

In the above worked example, the requirement for input impedance to be a particular value results in a unique set of design values for R_1 and R_2. In many practical cases the signal source has a negligibly small self impedance, as occurs for example when the signal source emanates from the output of a previous op. amp. circuit, and so the choice of R_1 is not critical. In these cases any reasonable pair of values for R_2 and R_1 with a ratio which results in the desired value for gain is satisfactory. Practical values of R_1 and R_2 are typically in the range 100 Ω to 10 MΩ.

The effect of the gain of practical op. amps. not being infinitely large can be found as follows. The relationship in Equation 6.4 for the circuit which surrounds the op. amp. stands, and so does the condition $i_a = 0$. However, $v_{dm} \neq 0$ and in fact is related to the output voltage by the expression $v_o = A \cdot v_{dm}$. That is

A is very large so v_{dm} is small compared with v_o, but not zero as assumed hitherto.

$$v_{dm} = \frac{1}{A} \cdot v_o \tag{6.6}$$

Substituting these conditions for i_a and v_{dm} into Equation 6.4 gives

$$\frac{v_i}{R_1} + \frac{v_o}{AR_1} = -\frac{v_o}{AR_2} - \frac{v_o}{R_2} \tag{6.7}$$

This is easily rearranged to give the overall voltage gain

$$A_v = \frac{v_o}{v_i} = -\frac{R_2}{R_1} \cdot \frac{A\dfrac{R_1}{R_1 + R_2}}{1 + A\dfrac{R_1}{R_1 + R_2}} \tag{6.8}$$

This expression allows the effect of finite *A* on overall circuit gain to be calculated. It can be seen that provided the condition

$$A\frac{R_1}{R_1 + R_2} >> 1$$

A high gain ratio R_2/R_1 means $R_1 << R_2$ and this can result in the condition $A\,R_1/(R_1+R_2) >> 1$ not being obeyed.

is maintained, the right-hand factor closely approximates to unity and the circuit gain is accurately given by the ratio $(-\ R_2/R_1)$. In practice, the circuit gain closely approximates this ideal ratio provided the desired gain is not too high.

Worked Example 6.2

Calculate the expected circuit gain for the circuit designed in worked Example 6.1 assuming the op. amp. to have a differential gain which is typical of the 741 type op. amp.

Solution: First,

$$A \frac{R_1}{R_1 + R_2} = 200\,000 \times \frac{1}{1 + 10} = 18\,182$$

Hence Equation 6.8 gives

$$A_v = \frac{v_o}{v_i} = -\frac{10}{1} \cdot \frac{18182}{1 + 18182} = -9.9995$$

In this example the circuit gain is affected hardly at all by the finite gain of the op. amp. This is because the op. amp. gain of 200 000 is very much larger than the magnitude of the desired circuit gain of -10. Using negative feedback terminology, the negative feedback is very strong (high loop gain) and so A_v is almost independent of A (see Chapter 2 concerning sensitivity to gain variations).

Now examine what effect the parameters of a practical op. amp. have on the output and input impedances of the inverting voltage amplifier. For the output impedance it is necessary to derive an expression for the self impedance Z_o of the Thevenin equivalent circuit which represents the output of the inverting voltage amplifier. In series with Z_o is the generator $A_v \cdot v_i$ which provides the inverted output voltage signal. By replacing the input signal by a short circuit, $v_i = 0$ the generator $A_v \cdot v_i$ disappears, thus leaving Z_o on its own in the Thevenin equivalent circuit. Thus, by connecting a hypothetical voltage source v at the output terminals, and determining the current i which flows, the output impedance Z_o is obtained from the ratio $Z_o = v/i$. Applying these constraints to the inverting voltage amplifier gives the circuit shown in Fig. 6.5. The op. amp. model is based on a voltage-controlled voltage source with gain constant $-A$. The op. amp. output impedance r_{oa} has been added to the model to determine its effect on the circuit output impedance Z_o. The op. amp. input impedance r_{ia} is also included for completeness.

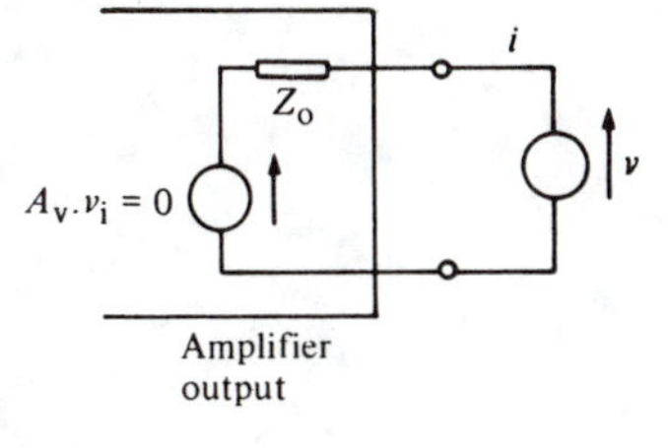

The gain constant in Fig. 6.5 is negative $-A$, because the signal is applied to the inverting input terminal of the op. amp.

The first step is to notice that the current i, caused by the hypothetical voltage generator v, splits in two parts,

$$i = i_1 + i_2 \tag{6.9}$$

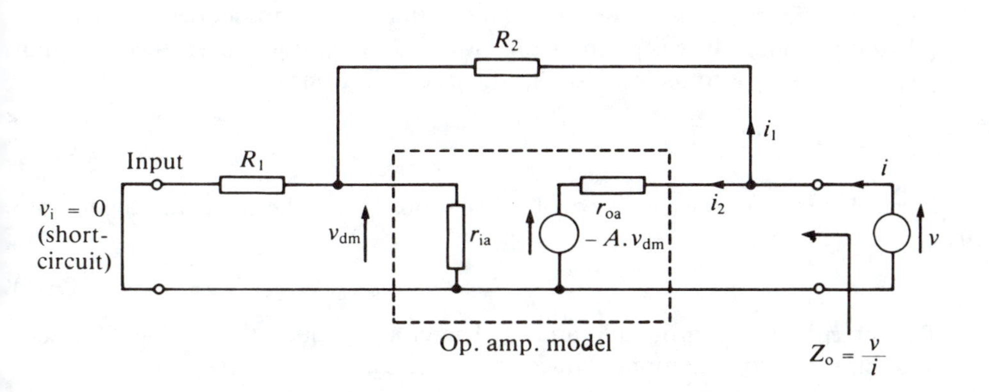

Fig. 6.5 Inverting voltage amplifier incorporating op. amp. model.

Current i_1 flows through R_2 and then through R_1 and r_{ia} which effectively are connected in parallel. Thus by Ohm's law

$$i_1 = \frac{v}{R_2 + R_1 \| r_{ia}} \tag{6.10}$$

Current i_2 flows through r_{oa}, and again by Ohm's law

Taking care with the minus signs.

$$i_2 = \frac{v - (-A \cdot v_{dm})}{r_{oa}} = \frac{v + A \cdot v_{dm}}{r_{oa}} \tag{6.11}$$

The voltage v_{dm} has been introduced into this expression, which has to be eliminated because the expression for circuit output impedance, $Z_o = v/i$, involves only the voltage v. This is achieved by noting that in the Fig. 6.5 v_{dm} is related to v by the potential divider formed by $(R_1 \| r_{ia})$ and R_2, thus

$$v_{dm} = \frac{R_1 \| r_{ia}}{R_2 + R_1 \| r_{ia}} \cdot v \tag{6.12}$$

Substituting for v_{dm} in Equation 6.11 and then substituting Equations 6.11 and 6.10 into Equation 6.9 gives

$$i = \frac{v}{R_2 + R_1 \| r_{ia}} + \frac{v}{r_{oa}} + \frac{A}{r_{oa}} \cdot \frac{R_1 \, r_{ia}}{R_2 + R_1 \| r_{ia}} \cdot v$$

This can be rearranged to give

$$\text{Output impedance, } Z_o = \frac{v}{i} = \frac{r_{oa}}{1 + \dfrac{r_{oa}}{R_2 + R_1 \| r_{ia}} + A \cdot \dfrac{R_1 \| r_{ia}}{R_2 + R_1 \| r_{ia}}} \tag{6.13}$$

This result is expected in view of the general effects of feedback on input and output impedance as described in Chapter 4.

It is important to notice here that r_{oa} is divided by a term which includes A, the op. amp. gain and so the output impedance of the circuit Z_o is very much less than the output impedance of the op. amp. Taking the previous worked example for which $R_1 = 1$ kΩ and $R_2 = 10$ kΩ, and using the typical values for the 741-type op. amp. in Table 6.1, Equation 6.13 gives $Z_o = 0.0041$ Ω. This is an extremely low value and is very close to the ideal value of $Z_o = 0$ Ω.

Finally, turn to the exact formulation of the input impedance. The inverting voltage amplifier circuit is shown in Fig. 6.6 with the op amp. model inserted. The required input impedance is given by the ratio $Z_i = v_i/i_1$. For convenience the input impedance looking into the right of R_1 is marked, Z_i'. Hence,

Comparing this expression with Equation 6.7 $Z_i = R_1$, shows that we can expect Z_i' to be very small.

$$Z_i = \frac{v_i}{i_1} = \frac{i_1 R_1 + v_{dm}}{i_1} = R_1 + Z_i' \tag{6.14}$$

It remains to find Z_i' which is equal to the ratio v_{dm}/i_1. The current i_1 divides into two parts i_2 and i_{ia},

$$i_1 = i_{ia} + i_2 \tag{6.15}$$

The current i_2 flows through R_2 and r_{oa}. Therefore, i_2 equals the voltage difference across this series combination, divided by $(R_2 + r_{oa})$,

Again being careful with the minus signs.

$$i_2 = \frac{v_{dm} - (-A \cdot v_{dm})}{R_2 + r_{oa}} = v_{dm} \cdot \frac{1 + A}{R_2 + r_{oa}} \tag{6.16}$$

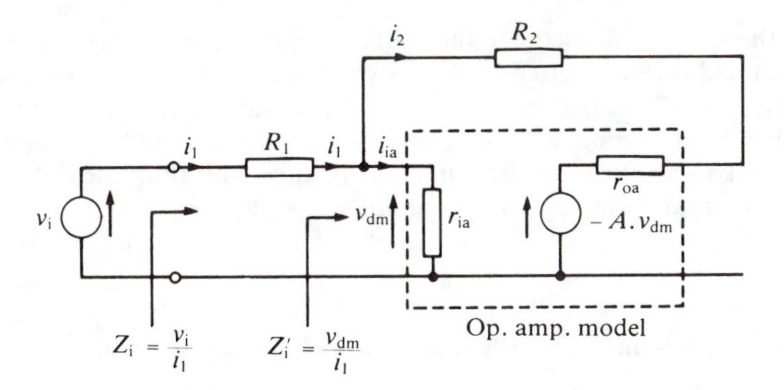

Fig. 6.6 Input impedance of inverting voltage amplifier circuit.

Also by Ohm's law

$$i_{ia} = \frac{v_{dm}}{r_{ia}} \qquad (6.17)$$

Substituting both Equations 6.16 and 6.17 into 6.15 gives

$$i_1 = v_{dm} \left(\frac{1}{r_{ia}} + \frac{1 + A}{R_2 + r_{oa}} \right)$$

From this $Z_i' = v_{dm}/i_1$ is obtained and when substituted in Equation 6.14 finally gives

$$\text{Circuit input impedance } Z_i = R_1 + \frac{1}{\left(\frac{1}{r_{ia}} + \frac{1 + A}{R_2 + r_{oa}} \right)} \qquad (6.18)$$

This expression is the same as that for the ideal op. amp. (that is, $Z_i = R_1$) with a term added. Equation 6.18 shows that provided A is large this term is quite small and it can often be neglected. For the worked example considered earlier, $R_1 = 1\ \text{k}\Omega$ and $R_2 = 10\ \text{k}\Omega$, and taking typical parameters for a 741-type op. amp., Equation 6.18 gives $Z_i = 1000.05\ \Omega$. This differs by only a tiny amount from the value of $Z_i = 1000\ \Omega$ obtained when the op. amp. is assumed to be ideal.

The conclusion which can be drawn from the foregoing analysis is that provided the op. amp. gain is very high the circuit behaviour is very close indeed to that obtained if the op. amp. is assumed to be ideal. This fact is generally true for all op. amp. negative feedback circuits. Together with the low cost of op. amps., it provides the reason for the very wide-spread use of these circuits. This is because the circuit performance is defined almost entirely by the circuit components surrounding the op. amp. and is accurately defined if these components have low tolerances.

Circuits Based on the Inverting Voltage Amplifier

Further op. amp. circuits which are derived from the inverting voltage amplifier circuit are discussed in this section.

The first is the *summing amplifier*. The existence of the virtual earth in the inverting voltage amplifier (node X in Fig. 6.4), means the function of R_1 can be

Although the virtual earth point has some of the attributes of an earth it is not identical to a true earth. For example, shorting the virtual earth point to actual earth would have no effect if the virtual earth was a proper earth, but in this case would remove the input voltage to the op. amp. and so stop the circuit working.

viewed as the conversion of the input voltage v_i into a current which is fed to the virtual earth and then converted back to a voltage at the circuit output equal to R_2 times the current. Obviously currents from other input voltages can be fed into the virtual earth in the same way, where they are summed and the total current converted back to a voltage v_o by R_2 and the op. amp. This principle is illustrated in the following worked example of a summing amplifier.

Worked Example 6.3 Describe a summing amplifier to sum three voltage signals v_A, v_B and v_C.

The extension of this idea to the summation of any number of inputs is obvious.

Solution: The summing amplifier is shown in Fig. 6.7. The current i into the virtual earth at X is given by

$$i = i_A + i_B + i_C$$

Because the voltage at X is virtually zero (assuming the op. amp. to be ideal) the currents i_A, i_B and i_C can be replaced by the Ohm's law relationships as follows.

$$i = \frac{v_A}{R_A} + \frac{v_B}{R_B} + \frac{v_C}{R_C} \tag{6.20}$$

The action of the feedback resistor R_2 and the op. amp. is to convert current i to a voltage at the circuit output,

This is because R and the op. amp. form a current-voltage converter.

$$v_o = -R_2 \cdot i$$

Substituting for i using Equation 6.21 gives

$$v_o = -\left(\frac{R_2}{R_A} \cdot v_A + \frac{R_2}{R_B} \cdot v_B + \frac{R_2}{R_C} \cdot v_C\right) \tag{6.21}$$

For example, if equal resistors are chosen, $R_A = R_B = R_C = R_1$ say

By choosing R_A, R_B, R_C to have unequal values v_O can be made equal to the sum of the input voltages, each scaled by its own individual factor.

$$v_o = -\frac{R_2}{R_1}(v_A + v_B + v_C) \tag{6.22}$$

Thus, the output is equal to the sum of the input voltages, scaled by the factor $(-R_2/R_1)$. The resistances R_1 and R_2 can be chosen to give any required scale factor.

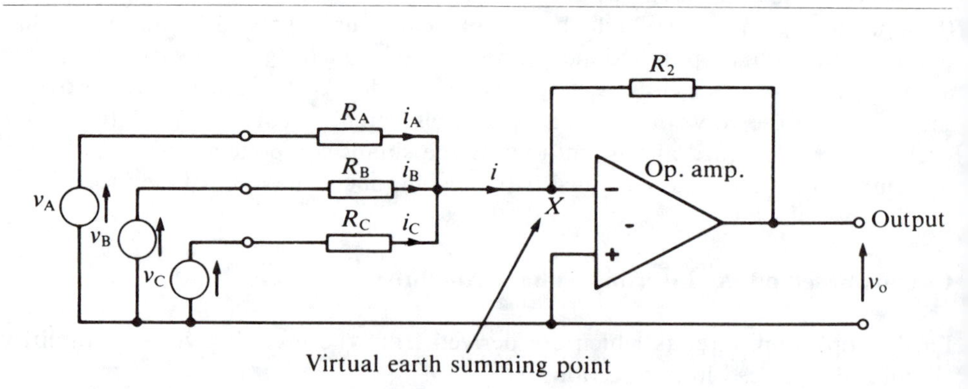

Fig. 6.7 Summing amplifier.

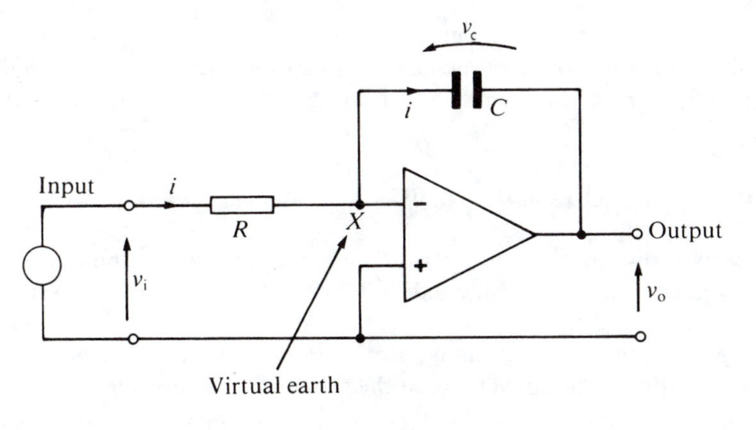

Fig. 6.8 Integrator circuit.

Other circuits can be obtained from the basic inverting voltage amplifier by using general impedances Z_1 and Z_2 which may contain reactive as well as resistive components in place of R_1 and R_2. The input-output relationship of Equation 6.5 is now written as

$$\text{Overall voltage gain, } A_v = -\frac{Z_2}{Z_1} \tag{6.23}$$

Various possibilities for Z_1 and Z_2 can be explored. One choice is to use a resistance for Z_1 and a capacitance for Z_2, as shown in Fig. 6.8, which acts as an *integrator circuit*. If the signals are assumed to be steady state sinusoids of angular velocity $\omega = 2\pi f$, the analysis using phasor notation and complex quantities is as follows.

$$Z_1 = R \text{ and } Z_2 = \frac{1}{j\omega C}$$

The complex notation is explained in any book on circuit theory.

Thus Equation 6.23 provides

$$\hat{A}_v = \frac{\hat{V}_o}{\hat{V}_i} = -\frac{1}{j\omega CR} \tag{6.24}$$

The ω in the denominator shows that the magnitude of the gain is inversely proportional to signal frequency, which is a characteristic of integrator behaviour.

That this circuit performs integration is made clearer if time domain signals are considered. The current into the virtual earth point is given, as before, by Ohm's law: $i(t) = v_i(t)/R$. This current flows through the capacitor. The capacitor current and capacitor voltage v_c are related by the well known basic relationship

To see this, consider the sinusoid $v = A \cos \omega t$. Then the integral

$$\int v.dt = \frac{A}{\omega} \sin \omega t$$

which has an amplitude which is inversely proportional to ω.

$$i(t) = C\frac{dv_c(t)}{dt} \tag{6.25}$$

Now, because one side of capacitor C is connected to virtual earth, the voltage across the capacitor is related to the circuit output voltage by $v_c = -v_o$. Substituting this in Equation 6.25, and integrating both sides gives

$$v_o(t) = -\frac{1}{C}\int i(t)\cdot dt + v_o(0) \tag{6.26}$$

If it is important to have $v_O = 0$ at $t = 0$ a switch (possibly electronically operated) can be connected across the capacitor and opened at time $t = 0$.

Hence, $v_o(0)$ is the constant of integration and reflects the fact that at time $t = 0$ the output voltage may be non-zero owing to the capacitor carrying some charge from an earlier excitation of the circuit. Substituting $i(t) = v_i(t)/R$ into Equation 6.26 gives the final result

Note this equation satisfies a dimensional check. The time constant CR has units of seconds while the integral of volts with respect to time has units of volts-seconds. After cancelling the seconds this leaves the right hand side having units of volts which agrees with the left hand side.

$$v_o(t) = -\frac{1}{CR}\int v_i(t)\cdot dt + v_o(0) \qquad (6.27)$$

This clearly shows that the output voltage is proportional to the time-integral of the input voltage (plus the initial condition $v_o(0)$). The scale factor is set by appropriate choice of R and C.

Op. amp. integrators have many uses, for example in analogue computers. Another application is the conversion of the signal from an accelerometer (which, of course, responds to the acceleration of some physical object to which it is attached) to a signal which indicates an object's velocity.

The integral of acceleration with respect to time is velocity.

Another circuit is obtained by exchanging R and C in the previous circuit. The result is the *differentiator circuit*.

Exercise 6.1 Show that the input-output relationship for the differentiator circuit is given in the time domain by

$$v_o(t) = -CR\cdot\frac{dv_i}{dt} \qquad (6.28)$$

Integration is the inverse operation to differentiation, so it is not surprising that in the circuit gain is now proportional to frequency, ω.

and for steady-state sinusoidal signals is given by

$$\hat{V}_o = -j\omega CR\cdot\hat{V}_i \qquad (6.29)$$

Potentially, there are many uses for differentiator circuits. However, a number of problems are present with analogue differentiators which limits their range of application. One problem occurs owing to the presence of random electrical noise in the input signal. The level of signal noise is usually small, but it is a fact of life of practical electrical circuits and that it can never be eliminated. Typical electrical noise has components which extend to quite high frequencies. Now Equation 6.29 shows that the gain of the differentiator circuit (indeed any differentiator circuit) is directly proportional to signal frequency. Thus, the higher frequency components of the unwanted noise are greatly emphasized by the differentiator. This problem can be partially corrected by limiting the gain at higher frequencies by, for example, connecting a capacitor across R to reduce its impedance at higher frequencies. However, this changes the behaviour of the circuit, which does not then act as an ideal differentiator.

A full treatment of this is given by Clayton, G.B. *Operational Amplifiers*, 2nd ed. (Butterworth, 1979) pp. 244 et seq.

Non-inverting Voltage Amplifier

Note the plus and minus signs on the op. amp. inputs are reversed compared with the inverting voltage amplifier. This is because for the feedback to be negative the feedback path R_2, R_1 must be connected to the inverting input terminal of the op. amp. It can also be seen that this circuit is an example of series-voltage feedback as described in Chapter 4.

The inverting voltage amplifier considered previously produces an inverted signal at the output. Sometimes a non-inverted output is needed. Where it is, a second inverting voltage amplifier could be connected to the output of the first to invert the signal back again to give an overall gain which has a positive sign. A simpler solution is to use the *non-inverting voltage amplifier* shown in Fig. 6.9. This circuit carries the further advantage of having a very high input impedance.

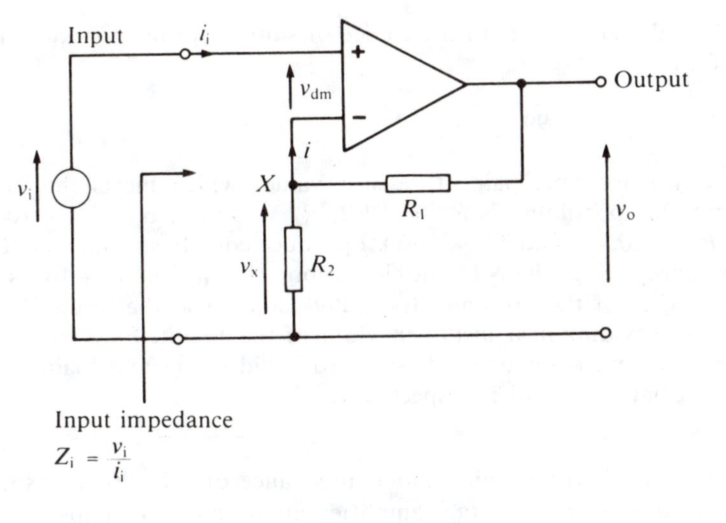

Fig. 6.9 Non-inverting voltage amplifier.

The non-inverting voltage amplifier circuit is recognized as one of the four basic feedback configurations described in Chapter 4. Hence the techniques of Chapters 4 and 5 can be used to analyse this circuit. However, it has been shown that in practice op. amp. circuit behaviour closely follows that indicated by assuming the op. amp. is ideal ($v_{dm} = 0$, $i_i = 0$) and can be analysed using basic circuit theory.

Thus, $v_i = v_x$, where v_x is the voltage produced by the potential divider formed by resistances R_1 and R_2. No current is drawn at node X because the op. amp. is assumed ideal, and so using the usual potential divider expression gives

$$v_i = v_x = \frac{R_2}{R_1 + R_2} \cdot v_o \tag{6.30}$$

The op. amp. output voltage sets itself so that this relation is satisfied. Therefore, the voltage gain is given by

$$A_v = \frac{v_o}{v_i} = 1 + \frac{R_1}{R_2} \tag{6.31}$$

Reference to Tables 4.2 and 4.3 shows that this circuit arrangement (series-voltage feedback) reduces the output impedance of a practical op. amp. Thus, for practical purposes

See Problem 4.5.

$$\text{Output impedance, } Z_o = 0\ \Omega \tag{6.32}$$

Most op. amp. circuits have parallel sensing feedback at the op. amp. output and so $Z_O \approx 0$. This point is important to remember.

The input impedance of the circuit is obtained as follows: Any input current which flows into the circuit, i_i, also has to flow through the input terminals of the op. amp.; so $i_i = i_{ia}$. However, for an ideal op. amp. $i_{ia} = 0$ so the current flowing into the circuit is also zero. Thus, since $i_i = 0$,

$$\text{Input impedance, } Z_i = \frac{v_i}{i_i} = \infty\ \Omega \tag{6.33}$$

Worked Example 6.4

Design an op. amp. non-inverting amplifier to provide a voltage gain of $+100$.

Solution: The design reduces to the choice of suitable values for R_1 and R_2. From Equation 6.31

$$\frac{R_1}{R_2} = A_v - 1 = 99$$

Hence, any reasonable pair of resistor values which meets this condition is satisfactory. One possibility is $R_2 = 1\ \text{k}\Omega$, $R_1 = 99\ \text{k}\Omega$, but $R_2 = 100\ \Omega$, $R_1 = 9.9\ \text{k}\Omega$ or $R_2 = 10\ \text{k}\Omega$, and $R_1 = 990\ \text{k}\Omega$ produce equally satisfactory results for a typical op. amp. such as the 741. The closed-loop circuit gain of + 100 is very much less than the gain of the op. amp. (typically 200,000) so the feedback is strongly negative thus providing near-ideal behaviour of the circuit. Extremely high or low resistance values must not be used, so as to avoid offset and loading effects, as explained in Chapters 7 and 5, respectively.

A detailed analysis of the non-inverting voltage amplifier, taking finite A, r_i, r_o into account is given by Ahmed, H. and Spreadbury, P.J. *Electronics for Engineers*, (Cambridge University Press, 1978) pp. 112 et seq.

The very high, almost infinite, input impedance of this circuit distinguishes it from the previous inverting voltage amplifier circuit, and is perhaps its most useful feature. When the input signal originates from the output of some other negative op. amp. circuit then the near-zero self impedance of the signal source usually means that a unity voltage coupling factor is obtained and the high-input impedance of the non-inverting amplifier, Fig. 6.9, is of no special value. However, when the input signal originates from some source having a significant self impedance (for example, a microphone or other transducer) to ensure a good coupling factor and therefore accurate amplification, an amplifier with high input impedance is essential.

See Equation 3.18 in Chapter 3.

Exercise 6.2 Show that the input resistance of the circuit in Fig. 6.9 in terms of the gain A and the input resistance r_i of the op. amp. is given by

$$Z_i = r_i \left(1 + \frac{A\,R_2}{R_1 + R_2}\right)$$

[Since A is large the second term in parenthesis usually dominates. Also, note that if there is a phase lag in A, which is frequently the case in practice, then the input impedance is complex (often capacitive).]

If the buffering effect owing to high-input impedance is required only and voltage gain is not important then choosing A_v to be unity allows a convenient simplification. Setting $A_v = 1$ in the gain relationship, Equation 6.31 gives the requirement

$$\frac{R_1}{R_2} = 0$$

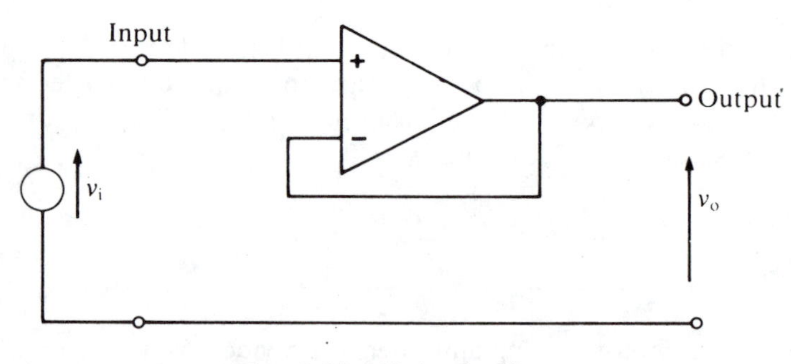

Fig. 6.10 Unity buffer.

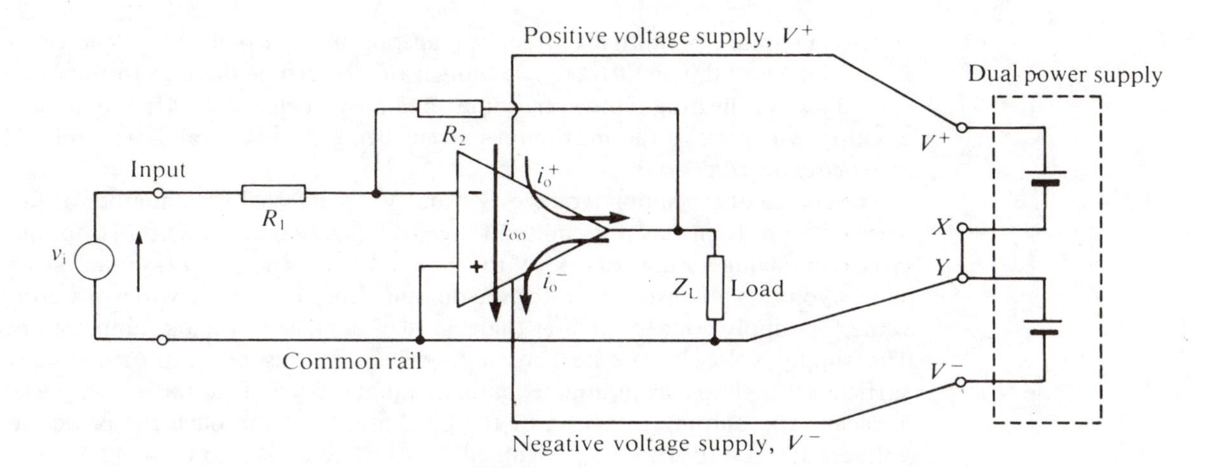

Fig. 6.11 Op. amp. power supply connections.

This condition is satisfied by setting $R_1 = \infty$ or $R_2 = 0$, or both. Choosing both, R_2 is removed from Fig. 6.9 and R_1 is replaced by a short-circuit. The result is shown in Fig. 6.10. This circuit is called the *unity buffer*. The circuit is attractive because it provides very high quality buffering without the need for accurate passive components.

Power Supply Connections to the Op. Amp.

Op. amps. like all active devices and circuits, require power to be supplied from an independent source. In most, if not all, op. amps. this is done by connecting a pair of positive and negative constant voltage sources, V^+ and V^-, to pins on the op. amp. The connections are indicated in Fig. 6.11. The op. amp. circuit illustrated is the inverting voltage amplifier, but the general principles are the same for any op. amp. circuit. The dual power supply is necessary and if two floating voltage sources are provided they must be tied down by connecting the negative terminal of the V^+ constant voltage source, to the positive terminal of the V^- constant voltage source. This is the link X–Y. The link is the zero-volts point and is connected to the common rail of the op. amp. circuit. It is important to remember to make this connection to the common rail of the op. amp. circuit. If not, the power supply although ensuring the correct voltage difference between V^+ and V^-, floats and V^+ and V^- adopt some arbitrary voltage with respect to the common rail and so prevent proper operation of the op. amp.

See Fig. 6.1b for the pin connections of the 741.

It can be seen that the op. amp. itself does not have a zero-volts point for connection to the common rail. At first sight such a connection might seem essential to allow a complete path for the load current to flow from the single-ended op. amp. output and through the load impedance to the common rail. In fact current paths exist without this connection. For instants in time where the op. amp. output current is positive the op. amp. supplies the current from the V^+ supply as indicated by the path i_o^+. For instants where the output current is negative the current flows back into the op. amp., via path i_o^-, and out to the V^- supply. The internal op. amp. circuitry also draws a small current for its own use from V^+ to V^- along path i_{oo}.

The impedance Z_L can be an actual load impedance, or the equivalent load formed by the input impedance to a following circuit.

The maximum permitted current for an op. amp. is typically in the region of 25 mA. Beyond this limit there is a danger of irreversible damage to the op. amp. caused by over heating. However, some op. amps., such as the 741, contain internal circuitry which stops the maximum current being exceeded, and are said to have *short-circuit protection*.

The choice of the supply voltages V^+ and V^- is affected by a number of factors. The op. amp. itself has a permitted range of supply voltages. The 741 op. amp. is typical in having a range of $\pm$ 3 V to $\pm$ 18 V for V^+ and V^-. Less than $\pm$ 3 V the supply voltages are insufficient for the op. amp. internal circuitry to work properly, while for supply voltages greater than $\pm$ 18 V damage to the op. amp. can occur. The supply voltages can be chosen anywhere in this range provided they are sufficiently high to accommodate that maximum excursions of the op. amp. signals. Typically the output voltage v_o of the op. amp. may approach the power supply voltages V^+, V^- to within 1 V. Thus if V^+, V^- are chosen to be $\pm$ 12 V, the peak op. amp. output voltage is limited to $\pm$ 11 V. If an attempt is made to drive the output signal outside this range, then distortion in the form of amplitude clipping occurs. The inverting and non-inverting signal inputs to the op. amp. also have a maximum working range. Like the op. amp. output range; this range is typically to within 1 V of the V^+ and V^- supplies. In op. amp. circuits such as the inverting voltage amplifier both inputs remain at approximately 0 V with respect to the common rail. However, in other circuits such as the non-inverting voltage amplifier or the unity buffer, both inputs although having negligible voltage differences can have appreciable voltages with respect to the common rail.

It is probably prudent to consider the limit to be $\pm$ 10 V to allow a margin.

In most circuits symmetrical power supplies are used; that is, $|V^+| = |V^-|$. This is because in very many applications the signals alternate symmetrically about the 0 V level. When the signals are not symmetrical there is no fundamental need to adopt symmetrical power supply voltages. For example, either V^+ or V^- can be made equal to 0 V since this allows a single voltage power supply to be used which is cheaper than a dual power supply. An example of this is the use of a unity buffer circuit to buffer a signal which is known to lie in the range 2 V to 5 V, say. In this case $V^- = 0$ V and $V^+ \geqslant 7$ V would prove satisfactory for a 741 type of op. amp., and the V^- pin of the op. amp. is connected directly to the zero volts common rail potential thus avoiding the need for a separate V^- power supply.

A treatment of further technicalities of single power supply operation is to be found in Clayton, G.B. *Operational Amplifiers*, 2nd ed. (Butterworth, 1979) p. 379 et seq.

Summary

The op. amp. is a very high gain differential amplifier. It is mass produced in integrated circuit form and therefore is inexpensive. The usefulness of the op. amp. comes from the basic property of feedback circuits that with large amounts of negative feedback the circuit performance is precisely defined by the feedback components.

Op. amp. circuits are analysed with good accuracy without using feedback theory by assuming the op. amp. to be ideal. The presence of an ideal op. amp. in an amplifier circuit constrains the current and the differential voltage at the op. amp. input terminals to be both zero. These constraints when included with the equations for the circuit surrounding the op. amp. permit the performance of the circuit to be readily analysed.

A basic and very useful op. amp. circuit is the inverting voltage amplifier. For this circuit

$$\text{Voltage gain} \approx -\frac{R_2}{R_1}$$

$$\text{Input impedance} \approx R_1\ \Omega$$

$$\text{Output impedance} \approx 0\ \Omega$$

The virtual earth in the circuit can be used to provide a summing property for more than one input. Also R_2 and R_1 can be generalised to impedances Z_2 and Z_1 and this allows circuits to be created which perform other operations on the input signal, such as integration with respect to time.

Another basic op. amp. circuit is the non-inverting voltage amplifier. For this circuit

$$\text{Voltage gain} \approx 1 + \frac{R_2}{R_1}$$

$$\text{Input impedance} \approx \infty\ \Omega$$

$$\text{Output impedance} \approx 0\ \Omega$$

This circuit provides amplification without inversion of the signal waveform. The circuit has the important property that the input impedance is very high and is therefore useful for buffering voltage signal sources with significant self-impedance. If voltage amplification is not required then by setting $R_1 = \infty$ and $R_2 = 0$, the very simple unity buffer circuit results which retains the buffering property but dispenses with external feedback resistors R_1 and R_2.

Dual power supplies are usually required for the op. amp. These must be correctly connected and be of appropriate value. The power supply voltages must lie in the working range of the op. amp. and be large enough to permit the maximum signal swings at the op. amp. inputs and output.

Problems

6.1 For Fig. 6.4, assuming the op. amp. is ideal,
  (i) calculate R_1 and R_2 to provide a circuit gain of -100, and input impedance of 10 kΩ;
  (ii) if $R_1 = 10$ kΩ and R_2 is a fixed resistor of 100 kΩ in series with a resistor which is variable over the range 0 to 50 kΩ calculate the range of variation for A_v and Z_i.

6.2 If $R_1 = 1$ kΩ and $R_2 = 20$ kΩ in parallel with a capacitor $C = 0.1$ μF calculate the magnitude of the voltage gain at signal frequencies of (i) 0 Hz and (ii) 2000 Hz.

6.3 A low-pass filter passes low frequencies and attenuates high frequencies. Show that the circuit in Problem 6.2 acts as a simple low-pass filter. (Remember a capacitor at 0 Hz has infinite impedance and this falls towards zero as the frequency increases towards high frequencies.) Relocate the capacitor to make a high-pass filter.

6.4 Given an op. amp. with $A = 100\,000$, $r_{ia} = 1$ MΩ, $r_{oa} = 100$ Ω, calculate the actual A_v, Z_i, Z_o for the circuit in Fig. 6.4 given that $R_1 = 100$ kΩ, $R_2 = 10$ MΩ.

6.5 Draw op. amp. circuits to perform the following functions

(i) $v_o = -10(v_A + v_B + v_C + v_D)$

(ii) $v_o = -10v_A - 20v_B - 5v_C$

(iii) $v_o = -v_A - 2v_B + 10v_C$

(Hint: Use two inverting voltage amplifier circuits for part (iii).) Assume the op. amps. are ideal and the largest resistor in each circuit has a value of 100 kΩ.

6.6 Calculate the values of components in a circuit to perform the operation $v_o = -10 \int v_i \cdot dt + v_o(0)$. The input resistance is to be 100 kΩ.

6.7 An integrator circuit, Fig. 6.8, uses $R = 1$ MΩ, C = 2 μF. Assume the op. amp. is ideal and the circuit output voltage is initially at 0 V. A 1 V positive step is applied to the input. Sketch the output waveform and calculate the time it takes to reach a magnitude of 10 V.

6.8 For the non-inverting voltage amplifier, Fig. 6.9, calculate $|A_v|$ for the following cases

(i) $R_2 = 1$ kΩ, $R_1 = 25$ kΩ.

(ii) $R_2 = 10$ kΩ in series with a 0.2 μF capacitor. The signal frequency is 100 Hz. $R_1 = 100$ kΩ.

(iii) $R_2 = 100$ Ω and $R_1 = 1$ kΩ in parallel with a 0.2 μF capacitor. The signal frequency is 100 Hz.

6.9 For Fig. 6.9, if the op. amp. is assumed to have $r_{ia} = \infty$, $r_{oa} = 0$, and finite A, use basic feedback theory (Chapter 4) to show that

$$A_v = \frac{A}{1 + A \cdot \frac{R_2}{R_1 + R_2}} = \left(1 + \frac{R_1}{R_2}\right) \cdot \frac{1}{1 + \frac{1}{A}\left(1 + \frac{R_1}{R_2}\right)}$$

6.10 An inverting voltage amplifier has $R_1 = 100$ Ω, $R_2 = 1$ kΩ, and a load resistor $r_L = 1$ kΩ connected between the op. amp. output and the common rail. The op. amp. passes a standing current $i_{oo} = 1$ mA (see Fig. 6.11) but is otherwise ideal. A constant input voltage of +1 V is applied and the voltage supplies V^+, V^- are ± 12 V. Calculate

(i) the current supplied by each voltage supply,

(ii) the total power delivered by both power supplies,

(iii) the power dissipated in the load resistor, and

(iv) the power dissipated in the op. amp.

(Remember current flows through R_1 and R_2.)

Op. Amp. Non-idealities 7

Objectives

- ☐ To analyse the effects of op. amp. input offset voltage V_{os}, and how to reduce or cancel these effects.
- ☐ To appreciate the effects of op. amp. input bias currents I_b^+, I_b^-.
- ☐ To apply bias current compensation and the importance of the input offset current I_{os}.
- ☐ To describe the effect of op. amp. frequency response on op. amp. circuit performance.
- ☐ To understand how the effects of slew-rate limiting can be avoided and also the meaning of full-power bandwidth.

The Importance of Op. Amp. Non-idealities

Some op. amp. non-idealities were considered in Chapter 6. A real op. amp. has finite gain and finite input impedance. It also has non-zero output impedance. However, it was shown that provided large amounts of negative feedback are present in the circuit where the op. amp. is used, these non-idealities produce extremely small effects. Additionally, there are other non-idealities which are sometimes important and cannot be eliminated by negative feedback. First, in high-precision circuits, effects which, although small in absolute terms, can cause significant deviations from the ideal circuit behaviour. Non-idealities such as *offset voltage* and *bias current* are significant here. Second, because the internal op. amp. circuitry is usually designed to give high gain, and so contains a large number of active devices, the speed of response of an op. amp. is generally rather limited. Consequently, consideration of op. amp. *frequency response* and *slew-rate* are important, and are considered later in this chapter.

Offset Voltages

An op. amp. is a high-gain differential amplifier and so if the differential input voltage v_{dm} is made zero by shorting together the input terminals, the output voltage ideally should also be zero. In practice, however, this is not so and large positive or negative voltage is usually present at the op. amp. output which is obviously undesirable. This voltage remains even if both input terminals are shorted down to the common rail to make the common mode input voltage zero. This proves the effect is not owing to any finite common-mode amplification in the op. amp.

Prove this for yourself by looking at the definition of A_{cm} in Chapter 3.

In fact, this effect is caused by inbalances in the quiescent conditions within the op. amp. The internal circuitry of an op. amp. is quite complex and inbalances can occur at various points in the circuit, for example, mismatched transistors in the differential input stage in Fig. 6.2.

The output voltage of the op. amp. could be set to zero by applying, between the differential input terminals, a voltage of appropriate magnitude and polarity. This voltage is called the *offset voltage* V_{os}. The maximum magnitude of V_{os} for particular op. amp. types is specified in manufacturers' data sheets. The 741-type, for example, has a maximum offset voltage of 5 mV. However, this is a relatively large value compared with V_{os} for more recently introduced op. amps. The above definition of offset voltage leads to the alternative models for offset behaviour in op. amps shown in Fig. 7.1. The polarities indicated by the symbol for V_{os} are a matter of convention. The actual polarity of V_{os} can be positive or negative depending on the particular op. amp. specimen.

This means the actual V_{os} for a particular specimen of 741 op. amp. lies in the range −5 mV to +5 mV.

Worked Example 7.1 A 741-type op. amp. has its input terminals shorted together. Estimate the maximumvoltage which might be expected at the op. amp. output. Assume A = 200,000 and V_{os} = 5 mV.

Solution: Referring to Fig. 7.1a or b, if the input terminals are shorted together, this results in an e.m.f. V_{os} at the input to the hypothetical offset-free amplifier. Thus,

$$V_{out} = A.\ V_{os} = 200\,000 \times 5\text{ mV} = 1000\text{ V}$$

Actually ±1000 V because V_{os} can have either polarity.

Because the value quoted in specifications for V_{os} is the maximum *magnitude* the above result indicates that the output owing to offsets can be expected to be somewhere in the range −1000 V to +1000 V. Of course this range is impossibly wide. As shown in the previous chapter the power supplies to the op. amp. constrain the output to be a lot less than this, typically −10 V to +10 V. Therefore, when zero voltage is applied to an op. amp. the offset effect usually cause the op. amp. to saturate at its most positive or negative limit. Taking these effects into account, the transfer characteristic shown in Fig. 6.3 is modified as shown in the margin, where in this example zero input voltage gives −10 V at the output.

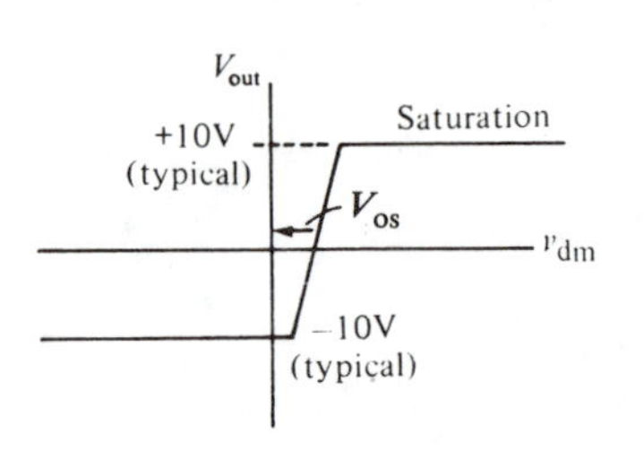

Apparently, therefore, there is a problem: How can the op. amp. be usefully used if voltage offsets cause such large unwanted output voltages? In fact, when negative feedback is applied to the op. amp., such as in the inverting and non-inverting amplifier circuits described in the previous chapter, the unwanted output voltage is very much reduced.

Once again negative feedback gives benefits.

It is important to be able to analyse the likely effects of the offset voltage while

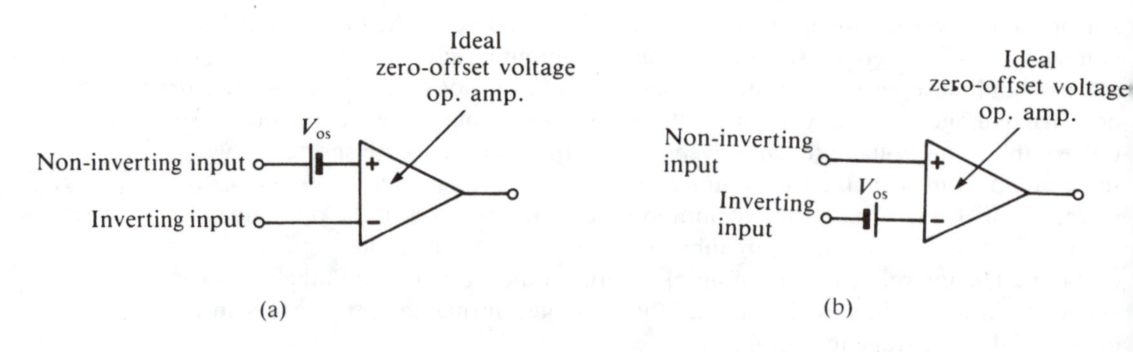

Fig. 7.1 Alternative models for op. amp. offset voltage.

still at the design stage. To do this either of the models for op. amp. offset voltage shown in Fig. 7.1 may be used. The following worked example illustrates this.

Worked Example 7.2

Derive a formula for the output voltage of the non-inverting voltage amplifier circuit owing to voltage offset effects. Estimate the maximum output offset voltage if a 741-type op. amp. is used and the circuit is designed to have a gain of 10. Assume the only op. amp. non-ideality present is voltage offset V_{os}.

Solution: The non-inverting voltage amplifier has been discussed in the previous chapter. See Fig. 6.10. To analyse this circuit the op. amp. is replaced by either of the voltage offset models in Fig. 7.1. Both these models are equivalent and so lead to identical results. In turns out that in this case the use of left-hand model allows the result to be written down by inspection. With this model the non-inverting voltage amplifier is as shown in Fig. 7.2. Note that V_{os} is in series with the input voltage, hence the input V'_{in} to the hypothetical offset-free op. amp. is given by

It can be worthwhile to try initially both models to see if a quick analysis suggests itself.

$$V'_{in} = V_{in} - V_{os} \tag{7.1}$$

Now the circuit to the right of V'_{in} is identical to the non-inverting voltage amplifier that has been analysed in the previous chapter. From this analysis and Equation 6.31

$$V_{out} = (1 + \frac{R_1}{R_2})V'_{in}$$

Thus, from Equation 7.1

$$V_{out} = (1 + \frac{R_1}{R_2}) \cdot V_{in} - (1 + \frac{R_1}{R_2}) \cdot V_{os} \tag{7.2}$$

Using the other model of course would give the same answer as this.

This shows that the output voltage is made up of the desired term,

$$(1 + \frac{R_1}{R_2})\, V_{in}$$

and an unwanted term,

$$-\,(1 + \frac{R_1}{R_2})\, V_{os},$$

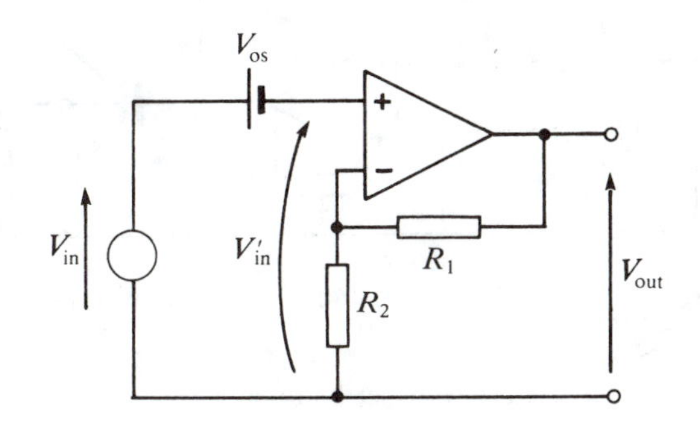

Fig. 7.2 Offset voltage in the non-inverting amplifier.

owing to voltage offsets.

For a gain of 10 then

$$(1 + \frac{R_1}{R_2}) = 10,$$

and therefore for the 741-type of op. amp. the maximum output offset voltage is given by

$$-(1 + \frac{R_1}{R_2}) \cdot V_{os} = -10 \times 5 \text{ mV} = -50 \text{ mV}$$

For a real circuit the actual output voltage owing to offsets lies in the range −50 mV to +50 mV, and the output voltage is not now saturated at ±10 V.

Exercise 7.1 For the inverting voltage amplifier circuit (Fig. 6.4) assume the op. amp. is ideal except for V_{os}, and show that

Hint: See the earlier analysis around Equation 6.4 and note that now $v_{dm} = -V_{os}$.

$$V_{out} = -\frac{R_2}{R_1} \cdot V_{in} - (1 + \frac{R_2}{R_1}) \cdot V_{os} \qquad (7.3)$$

Also show that for a 741-type op. amp. used in a circuit with a gain of −10, the unwanted output offset voltage lies in the range −55 mV to +55 mV.

Although in a negative feedback op. amp. circuit the unwanted output offset voltage is usually quite small, this voltage cannot always be neglected. This depends on the application. If, for example, high precision operation is required, even a small amount of output offset voltage may cause the circuit output to deviate from the ideal by an unacceptable amount. Difficulties can also arise if the input signal to the circuit is small (compared with V_{os}) and large amplification is required. The large resistor ratios needed to achieve this, can result in a large output offset voltage. For the non-inverting voltage amplifier the voltage gain is given by $(1 + R_1/R_2)$ and this also multiplies V_{os} in the right-hand term of Equation 7.2. Similarly, for the inverting voltage amplifier the magnitude of the voltage gain is R_2/R_1 which is also contained in the factor which multiplies V_{os} in the right hand term of Equation 7.3.

See Equation 7.2.

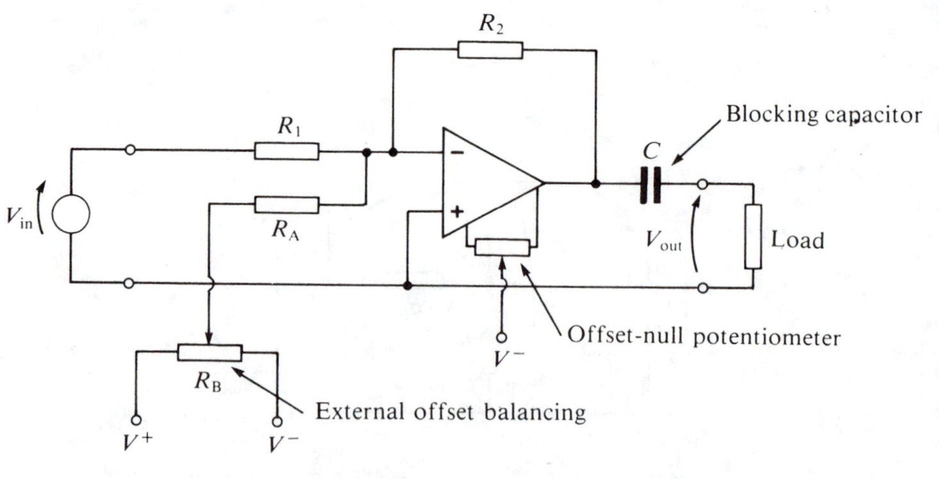

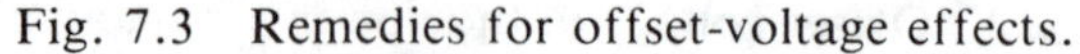

Fig. 7.3 Remedies for offset-voltage effects.

When the output offset voltage cannot be neglected a number of remedies are available. There are four principal remedies and three of those are illustrated in Fig. 7.3 for the inverting voltage amplifier. Usually only one of the techniques is employed. The four remedies are as follows.

Blocking Capacitor

This is the capacitor C. Because the capacitor has infinite impedance at zero frequency, the blocking capacitor prevents any constant offset voltage at the op. amp. output from reaching the load. The blocking capacitor of course also prevents any zero-frequency signal voltage from reaching the load, and so this method can only be applied where the signals are alternating with a frequency greater than some prescribed minimum. If the signals contain zero frequency components, then capacitor C should not be inserted. One of the following methods may be used.

Capacitor C introduces a low frequency effect in the voltage coupling,

Offset-Null Potentiometer

Many op. amps. are provided with arrangements for balancing out (that is, *nulling*) the offset voltage. In the 741-type a 10 kΩ potentiometer is connected between pins 1 and 5, and the sliding contact of the potentiometer is connected to a negative voltage supply, V . The same negative supply as the main one used for the op. amp. itself can be used. The inbalances inside the op. amp. are corrected by adjusting the potentiometer control until, with $V_{in} = 0$, an output of $V_{out} = 0$ is observed.

External Offset Balancing

The idea behind this method is to inject a voltage or current into the circuit to produce a component of output voltage which is equal and opposite to the unwanted offset component, thus cancelling out the latter. A variety of ways to do this exists. The main requirement for any chosen way is that the method of introducing the offset balancing voltage or current should not interfere with the normal signal amplifying properties of the op. amp. circuit. The arrangement shown on resistor R_A and potentiometer R_B shown in Fig. 7.3 achieves this. The potentiometer R_B is connected across the op. amp. voltage supplies V and V and allows the voltage applied to R_A to be adjusted to any value in this range. The other end of R_A is connected to the virtual earth of the op. amp. circuit and because of the summing property (see the previous chapter) the potentiometer voltage, scaled by the factor R_2/R_A, appears at the op. amp. output and can be adjusted to cancel the unwanted offset voltage. Making this factor small, say 1/100, lessens the sensitivity of the output voltage setting to the position of R_B.

Some other ways are to be found in Clayton G.B. *Operational Amplifiers*, 2nd ed. (Butterworth, 1979) p. 357 et seq.

Check for yourself the scale factor is R_2/R_A.

Lower Specified Maximum V_{os}

This, the last remedy to be mentioned, is perhaps the most obvious. With a maximum V_{os} of 5 mV the general-purpose 741 op. amp. has only moderately good offset voltage characteristics. Other op. amp. types are designed and manufactured (using laser trimming techniques, for example) to provide much lower V_{os} values. Best of all are the *chopper-stabilized* or *commutating* types which have V_{os} values in the region of a few microvolts. Of course, improved specifications are usually obtained for an increase in purchase price. However, this cost is offset to a certain extent because the need for extra passive components and precise circuit adjustments, used in the other remedies, are no longer needed.

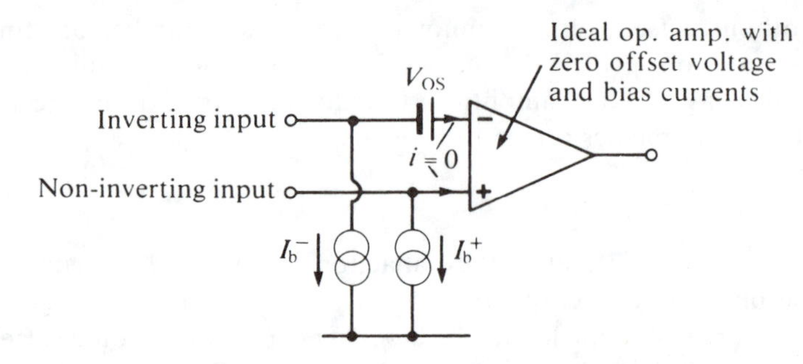

Fig. 7.4 Modelling of bias currents and offset voltage in an op. amp.

Bias Currents and Offset Currents

Commonly NPN transistors are used at the input stage, so I_b flows into the op. amp. If PNP transistors are used current flows out so I_b^+, I_b^- are then negative quantities.

The bipolar junction transistor differential amplifier circuit at the input stage of many op. amps. requires significant d.c. base currents to be drawn into the op. amp. input terminals to provide biasing for proper circuit operation. The current into the non-inverting input terminal is labelled I_b^+ and that into the inverting input terminal is I_b^-. These bias currents flowing through resistances in the circuit generate unwanted d.c. voltages. The general purpose 741 op. amp. has a bipolar-junction-transistor input stage and the maximum bias currents are specified at 0.5 μA at 25 °C. Where lower bias currents are needed, field-effect transistors are used for the input stage; a typical value of I_b is then 10 pA.

The bias currents can be modelled by two current generators across the inputs of a hypothetical op. amp. which is assumed to have no bias currents. The complete model for bias currents and offset voltage is shown in Fig. 7.4. The following worked example illustrates the use of this model and shows how the unwanted effects of bias currents can be compensated.

Worked Example 7.3

Analyse the effects of op. amp. bias currents and offset voltage on the output of the inverting voltage amplifier. Show how bias currents can be compensated.

This figure is used later to analyse the integrator circuit, with R_2 replaced by capacitor C.

Solution: The inverting voltage amplifier (see Fig. 6.4) is shown redrawn in Fig. 7.5 with the op. amp. model for bias currents and voltage offset. Note that resistor R has been added to the circuit. It is shown that this resistance can be used to compensate for bias-current effects, because the current I_b^+ flowing through it generates a voltage at node 1 that can be made equal to the voltage generated at node 3 by the current I_b^-, with no unwanted voltage at the output.

This circuit is easier to analyse than might at first appear. The analysis starts by determining the voltage V_1 at node 1 with respect to the common rail. To do this, first note that all of the current into generator I_b^+ must flow via R, because no current can flow along the path to the positive terminal of the hypothetical op. amp. since it is ideal and draws no current. Thus by Ohm's law $V_1 = -R.I_b^+$. Next, because the ideal op. amp. has zero voltage difference between its input terminals the voltage at node 2 with respect to the common rail is given by $V_2 = V_1$. Between nodes 2 and 3 is the voltage offset generator and so $V_3 = V_2 - V_{os}$. Combining these three equations gives

Why is the minus sign present in this equation?

Voltage V_3 corresponds to the virtual earth. This equation shows that in practice the point is not exactly at zero volts.

$$V_3 = -V_{os} - R.I_b^+ \tag{7.4}$$

The plus and minus signs on the right-hand term occur because the difference in bias currents may be of either polarity. Although the existence of I_{os} leads to an unwanted output component, this component is less than if no bias current compensation is applied.

Before leaving the topic of offsets we now examine their effects in the op. amp. integrator circuit. This is an important circuit and it turns out that offset effects, in the form of drift, can be significant.

As a preliminary, note that the inverting voltage amplifier and the integrator circuit differ only in the type of feedback component. Hence, the analysis of the integrator circuit can be done using Fig. 7.5 provided R_2 is replaced by C, as indicated. Because of the similarity of the two circuits much of the earlier analysis can be used for the integrator circuit. For the integrator circuit the current I_2 through the capacitor is related to the voltages across the capacitor by

$$I_2 = C\frac{\mathrm{d}}{\mathrm{d}t}(V_3 - V_0) = C\frac{\mathrm{d}V_3}{\mathrm{d}t} - C\frac{\mathrm{d}V_0}{\mathrm{d}t} \qquad (7.11)$$

Hence Equations 7.5 and 7.6 now become

$$\frac{V_i - V_3}{R_1} = C\frac{\mathrm{d}V_3}{\mathrm{d}t} - C\frac{\mathrm{d}V_0}{\mathrm{d}t} + I_b^- \qquad (7.12)$$

Substituting for V_3 using Equation 7.4 clears the variable from this expression, and after a little manipulation;

$$\frac{\mathrm{d}V_0}{\mathrm{d}t} = -\frac{V_i}{R_1C} - \frac{\mathrm{d}V_{os}}{\mathrm{d}t} - R\cdot\frac{\mathrm{d}I_b^+}{\mathrm{d}t} + \frac{1}{CR_1}(-V_{os} + R_1 I_b^- - R I_b^+) \qquad (7.13)$$

Because V_{os} and I_b^+ are constant the terms in Equation 7.13 involving their derivatives are zero and are deleted. After this, integrating both sides of the expression gives the final result:

$$V_0 = -\frac{1}{CR_1}\int V_i \cdot \mathrm{d}t + V_0(0) + \frac{t}{CR_1}(-V_{os} + R_1 I_b^- - R I_b^+) \qquad (7.14)$$

$V_0(0)$ is the usual constant of integration. It is the integrator output at time $t = 0$, due to any initial charge on the capacitor.

This expression is the same as that for the ideal integrator (see Equation 6.27) but with the addition of an unwanted right-hand term brought about by the existence of V_{os}, I_b^- and I_b^+. This unwanted term is fundamentally different from those of other circuits considered so far as it contains the time variable t. Thus, the error component continually increases with time at a constant rate until limited by the maximum positive or negative output voltage swing for the op. amp. This effect is called *drift*. Examination of Equation 7.14 shows that the rate of drift can be reduced by the reduction of V_{os} (by nulling, for example) and by using an op. amp. with low bias currents. Also bias current compensation can be applied, as in the exercise below.

It can increase in a negative or positive direction depending on circuit and op. amp. parameters. Why?

Physically, drift occurs because a constant current component of I_2 must be drawn through the feedback component to satisfy the circuit conditions at the op. amp. input owing to bias currents and V_{os}. This is supplied by the op. amp. output. If the feedback path allows non-fluctuating currents to flow in the steady state (as in the non-inverting voltage amplifier where the feedback component is a resistor) then a constant offset voltage results at the op. amp. output. However, in the case of the integrator circuit there is no steady state path through the capacitor and in forcing a constant current through it the op. amp. causes the capacitor to charge up as time passes.

Exercise 7.3 Bias current effects in the op. amp. integrator circuit can be compensated for by choosing the resistance R at the non-inverting op. amp. input to satisfy the condition $R = R_1$. Prove this by (i) considering the unwanted term in Equation 7.14, and also (ii) by using the general rule discussed earlier that for bias current compensation the resistances seen by each op. amp. input terminal should be equal.

Note the *resistance* of a capacitor at zero frequency is infinity.

By careful trimming and choice of op. amp. parameters, drift can be reduced or even apparently stopped at one moment in time. However, absolute elimination is not possible and in general eventually the integrator output drifts until the op. amp. output saturates, so rendering the circuit inoperative.

Integrator drift in applications such as analogue computers is dealt with by initializing the integrator output to have zero output voltage by discharging the capacitor voltage by means of a switch which is opened when computation starts. The circuit is then run for a period of time until the drift error exceeds some specified amount.

More details can be found in Clayton, G.B. *Operational Amplifiers*, 2nd ed (Butterworth, 1979) p. 225 et seq.

In some other applications, which typically contain several op. amp. circuits, integrator drift does not occur because a d.c. negative feedback path exists from the integrator output via other circuits to the integrator input. This acts in a similar way to the feedback resistor R_2 in the non-inverting voltage amplifier, and instead of drift a constant offset voltage appears at the integrator output.

Op. Amp. Frequency Response

Although the op. amp. has very high differential gain, this depends very much on the signal frequency. At quite modest frequencies the gain usually begins to fall off. The frequency response for the 741 type op. amp. is typical and the characteristic provided by manufacturers is shown in Fig. 7.6. The gain of the op. amp. has a high value, typically 200 000 (106 dB), at zero frequency. However, as the signal frequency increases to the region of 10 Hz a capacitor within the op. amp. causes the gain to fall. This is the *dominant* high-frequency effect. The fall in gain continues at 20 dB/decade for many decades of signal frequency until the next high-frequency

Vertical and horizontal axes of Fig. 7.6 have logarithmic scales.

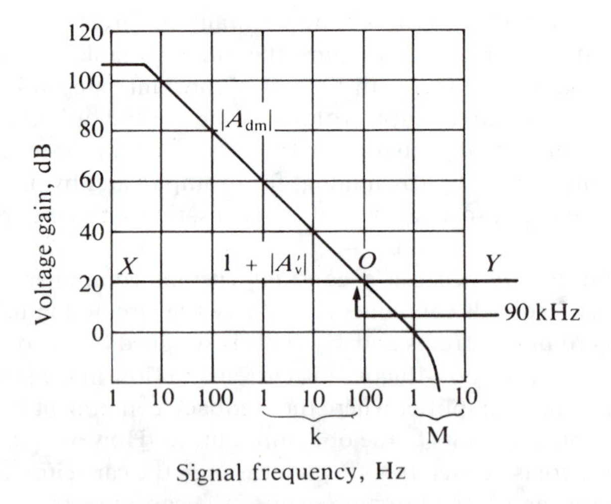

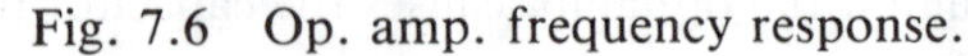
Fig. 7.6 Op. amp. frequency response.

effect comes into play in the region of 1 MHz. After this, the fall off in gain accelerates.

Although the first high-frequency effect occurs at quite a low frequency (in the region of 10 Hz) the usable signal bandwidth of most circuits using the op. amp. is considerably wider than this. The reason is that gain is traded off for bandwidth when negative feedback is applied. Calculation of the resulting bandwidth can be done using the analytic techniques described in Chapter 5. However, it turns out that the usable frequency range can be estimated directly from the published frequency-response curve for the op. amp. Example 7.4 shows how to do this.

Frequency-response effects are discussed in Chapter 3. See in particular Fig. 3.7. The op. amp. has no low-frequency effects and its amplification extends down to zero frequency.

Worked Example 7.4

An inverting voltage amplifier is designed to have a closed-loop gain of -10 at low frequencies. A 741-type of op. amp. is used. Estimate the closed-loop bandwidths of the circuit.

Solution: From the previous chapter (Equation 6.8) the closed-loop gain is given by

$$A_V = -\frac{R_2}{R_1} \cdot \frac{A\dfrac{R_1}{R_1 + R_2}}{1 + A\dfrac{R_1}{R_1 + R_2}} \tag{7.15}$$

and $A_V \approx -R_2/R_1 = A'_V$ say provided the following condition is satisfied (7.16)

$$|A|\frac{R_1}{R_1 + R_2} >> 1 \tag{7.17}$$

Because A generally has a phase angle at signal frequencies, the magnitude of the gain $|A|$ is used.

The fall in $|A|$ as signal frequencies are increased means that eventually the above condition is not satisfied. The circuit gain A_V then ceases to be well defined by the resistor ratio R_2/R_1, and this marks the limit of useful frequency range of the circuit.

Rearranging Equation 7.17 gives

$$|A| >> 1 + \frac{R_2}{R_1} = 1 + |A'_V| \tag{7.18}$$

The left-hand side of this inequality is the magnitude of the differential gain, $|A|$ and its variation with frequency is given by the manufacturer (see Fig. 7.6). The right-hand side of the inequality is defined by the desired circuit gain A'_V and so, $1 + |A'_V| = 1 + 10 = 11$. Expressed in decibels this is 20.8 dB. This quantity is given by the resistor ratio $(1 + R_2/R_1)$ and therefore is constant with respect to variations in frequency. It is shown as line XOY on Fig. 7.6.

Verify this figure of 20.8 dB.

For frequencies in the region XO, $|A|$ is greater than $1 + |A'_V|$. Reference to Equations 7.15 and 7.17 shows that the wider the gap the closer the circuit gain is to the ratio $-R_2/R_1$. For frequencies in the region OY, $|A|$ is less than $1 + |A'_V|$, the second term in the denominator of Equation 7.15 becomes less than unity and the gain A_V tends towards the expression $A\,R_2/(R_1 + R_2)$ and therefore falls off with increasing frequency as A falls off, and is no longer given by $-R_2/R_1$.

Thus the crossover point O represents the transition from a constant gain to a gain that falls off in proportion to A and so represents the limit of the useful range of operation. Now calculate the gain of the amplifier at the crossover frequency.

At the crossover point O, $|A| = 1 + |A'_V| = 11$. In addition, from the discussion of frequency-response effects in Chapter 3, because the crossover point occurs at a significantly higher frequency than the dominant frequency at 10 Hz, the phase angle of A is very close to $-90°$. Hence, using the j-notation the gain of A at the crossover point is equal to $-j11$. Now evaluate the closed-loop gain at this point.

Substituting this value for A into Equation 7.15

We have used here the fact that if $R_2/R_1 = 10$ then

$$\frac{R_1}{R_1 + R_2} = \frac{1}{11}$$

$$A_V = -(10) \cdot \frac{j11 \times \frac{1}{11}}{1 + j11 \times \frac{1}{11}} = \frac{10}{-1 - j}$$

and

$$|A_V| = \frac{10}{\sqrt{[(-1)^2 + (-1)^2]}} = \frac{10}{\sqrt{2}}$$

The closed-loop gain at the crossover point is thus reduced by the factor $\sqrt{2}$, and therefore point O corresponds to the upper half-power frequency of the closed-loop amplifier. Fig. 7.6 indicates this to be at about 90 kHz. The op. amp. gain remains high down to 0 Hz and so the closed-loop lower half-power frequency is also at 0 Hz. Therefore, the closed-loop bandwidth is given by

$$BW = f_U - f_L = 90 \text{ kHz} - 0 \text{ Hz} = 90 \text{ kHz}$$

Examining the effect of op. amp. frequency responses using a Bode diagram as in Fig. 7.6 is convenient because it allows the interaction op. amp. performance and expected closed-loop performance to be readily appreciated.

A good introductory treatment of frequency compensation is given by Clayton, G.B. *Operational Amplifiers*, 2nd ed (Butterworth, 1979) p. 43 et seq.

Bode plots are very useful for a number of other purposes such as the application of *frequency compensation* techniques to shape the op. amp. frequency response to obtain wider frequency range of operation or to avoid instability. The manufacturers of the 741 op. amp. have designed it to have internal frequency compensation. The first, or dominant frequency effect is purposely made very low, (at about 10 Hz) so that the op. amp. gain falls to unity (0 dB) at the onset of the next high-frequency effect (at about 1 MHz). Thus, over the whole of the useful amplification range the phase shift of the op. amp. is less than 180°, and so for a wide range of circuit applications the 741 op. amp. is free from instability problems. (This enhances the general applicability of the op. amp. but means the frequency range obtainable is not as wide as with some other op. amp. types.)

From Chapter 3 we know that if the total phase shift of the loop gain does not exceed 180% then it cannot encircle the critical point $-1 + j0$.

Slew Rate and Full-Power Bandwidth

If a differential input is applied to an op. amp., and this input steps suddenly from zero volts to some constant voltage the output voltage changes rapidly but not instantaneously at a rate called the *slew rate*. It is found that the rate of change of output does *not* depend on the magnitude of the input voltage step. Hence slew rate represents the maximum rate of change that is available at the op. amp. output, and is measured in volts/second.

The reason for the slew rate limit lies within the op. amp. For the op. amp. output voltage to change, the voltages must also be changed across any of the small capacitances which lie on the signal path inside the op. amp. The basic current-

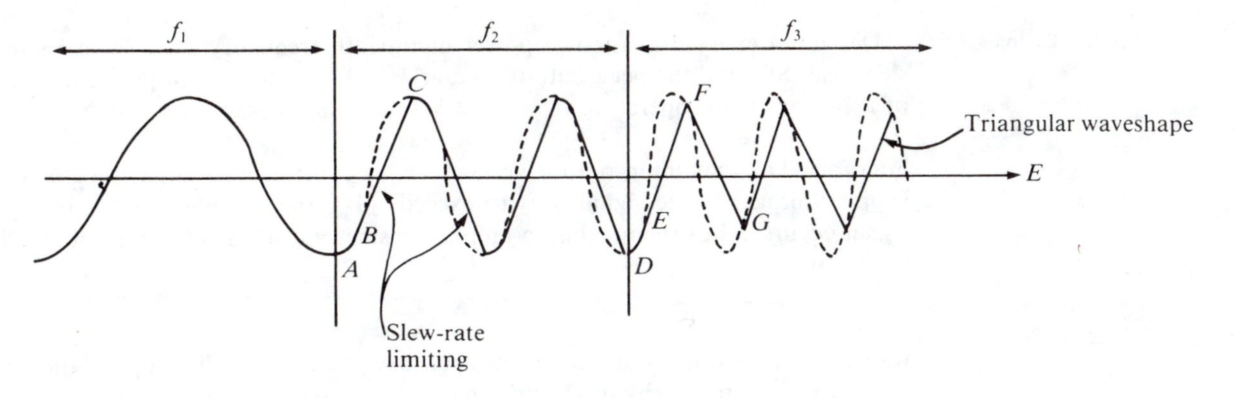

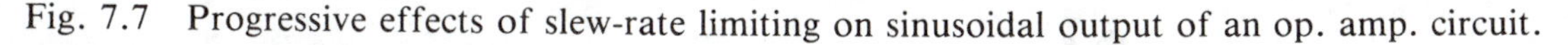

Fig. 7.7 Progressive effects of slew-rate limiting on sinusoidal output of an op. amp. circuit.

voltage relationship for a capacitor, $i = C(\mathrm{d}v/\mathrm{d}t)$, shows that to obtain a high rate of change in voltage $\mathrm{d}v/\mathrm{d}t$ high currents must be available to charge the capacitances up and down. In any practical circuit there must be some upper limit on the available currents and this explains why the slew rate limit exists.

The slew rate for a typical 741 op. amp. is 500 000 V/s. This is quite a large value and so for convenience slew-rate is usually expressed in volts per microsecond; the slew rate for the 741 is therefore 0.5 V/μs.

To see what can happen when slew rate limiting occurs consider as an example the inverting voltage amplifier when it is driven to produce a sinewave output. Fig. 7.7 shows the output voltage for three signal frequencies. For the low frequency f_1 the output waveform is nowhere steeper than the slew rate and there is no problem. At point A the frequency is increased to f_2. The slope of the waveform is low at this point but increases with time until at point B the slope of the sinewave is equal to the slew rate. Slew rate limiting then occurs. The actual output waveform must follow a straight line path which differs from the desired sinusoidal path (shown as a broken line). This condition continues until point C, where the desired sinewave coincides with the slew rate limited straight line. At this point control is regained by the op. amp. This behaviour occurs in a similar way on the following downward slope, and so on for every half cycle while the frequency remains at f_2. At point D the frequency is increased to f_3. Slew rate limiting again occurs at point E and continues to point F. However, at point F the sinewave has a slope which exceeds the slew rate and so slew rate limiting occurs immediately thus taking the op. amp. output along a straight line to G.

In practice slew rate for positive slopes need not be exactly the same as for negative slopes, but is usually approximately so.

The effect of slew rate limiting therefore is to take pieces out of the desired output waveform. For many applications this cannot be tolerated.

In the limit a repetitive output signal of any waveshape ends up as a triangular waveform. Can you prove this?

For any op. amp. circuit the following simple rule applies; the circuit is free from slew rate effects provided the instantaneous rate of change (i.e. slope) of the op. amp. output waveform is always less in magnitude than the slew rate.

Simple rule for freedom from slew rate effects.

In many cases the desired output signal is sinusoidal and a useful parameter for an op. amp. is the *full-power bandwidth*. This is defined as the greatest frequency of a full-voltage sinewave that can be output from the op. amp. without incurring slew rate effects.

Worked Example 7.5

Derive an expression for full power bandwidth frequency f_{FPBW} in terms of the slew rate, SR, and the peak output voltage V_{om}. Estimate the full-power bandwidth of a 741 op. amp. operating from ± 12 V power supplies.

Solution: The maximum output voltage V_{om} is controlled by the choice of power supply voltage. If the signal tries to exceed $\pm V_{om}$ then amplitude clipping of the signal occurs. Thus the maximum amplitude sinewave at the op. amp. output is of the form

$$v_o = V_{om} \cdot \sin(2\pi ft + \phi)$$

where f is the frequency and ϕ is some arbitrary phase angle. The slope of the output waveform is obtained by differentiating this expression to give

$$\frac{dv_o}{dt} = (2\pi f)\ V_{om} \cos(2\pi ft + \phi)$$

The points of maximum slope correspond to the zero crossing points of the v_O sinusoid.

The derivative varies with time and the greatest magnitude of slope occurs at those instants when $\cos(2\pi ft + \phi) = +1$ or -1. Thus

$$\left|\frac{dv_o}{dt}\right|_{max} = (2\pi f)\ V_{om}\ |\pm 1| = 2\pi f.\ V_{om} \tag{7.19}$$

At the full-power bandwidth frequency $f = f_{FPBW}$, the slope of v_o nowhere exceeds the slew rate but at one or more points just equals it. Thus from Equation 7.19

$$SR = \left|\frac{dv_o}{dt}\right|_{max} \text{(at frequency } f_{FPBW}) = 2\pi f_{FPBW} \cdot V_{om}$$

Hence, the desired expression is

$$f_{FPBW} = \frac{1}{2\pi}\frac{SR}{V_{om}} \tag{7.20}$$

Finally, for the 741 op. amp. the output can work typically to within 1 V of the power supplies, assumed here to be ± 12 V. So $V_{om} = 11$ V. The slew rate is typically 0.5 V/μs. Equation 7.20 therefore gives

Remember to include the factor 10^6 if slew rate is expressed in units of volts per microsecond.

$$f_{FPBW} = \frac{1}{2\pi} \times \frac{(0.5 \times 10^6)}{11} = 7234 \text{ Hz} \tag{7.21}$$

An important point emerges from this worked example. Note that for full sized output sinewaves the maximum signal bandwidth of 7234 Hz is considerably less than the figure of 90 kHz indicated for the op. amp. circuit in Worked Example 7.4.

In many, but not all, practical cases the usable frequency range is constrained more severely by slew rate limiting than by the signal bandwidth calculated from the op. amp frequency response.

By no means all op. amp. non-idealities have been considered here. A wider treatment is given by Clayton, G.B. *Operational Amplifiers*, 2nd ed (Butterworth, 1979). This book also gives a good discussion of how to measure op. amp. parameters.

Summary

Some of the more important op. amp. non-idealities have been considered and their effects on circuit performance have been analysed.

Offset voltage V_{os} and *bias currents* I_b^+ and I_b^- cause an unwanted offset voltage at the output of many op. amp. circuits. In the case of the op. amp. integrator circuit the effect is particularly significant because it takes the form of output drift. Several ways have been described to cancel the effects of V_{os}, including the use of an offset-null potentiometer. Bias currents effects can be compensated by choosing the resistances seen by both op. amp. input terminals to be equal. The effectiveness of this compensation depends on the value of *offset-current*, $|I_b^+ - I_b^-|$.

The useful signal frequency range of an op. amp. circuit is determined by the op. amp. *frequency response.* A Bode diagram is convenient for the examination of this.

Another effect is due to *slew rate* which limits the slope of the op. amp. output waveform. *Full-power bandwidth* provides a useful measure of the maximum permitted frequency of a full amplitude output sinewave if slew rate distortion is to be avoided. Often the full-power bandwidth is less than bandwidth indicated by frequency-response analysis.

Problems

7.1 An inverting voltage amplifier circuit comprises an op. amp. and resistors R_1 = 1 kΩ and R_2 = 100 kΩ. Calculate the maximum unwanted component of output voltage for the following cases
  (i) V_{os} = 1 mV, other non-idealities neglected,
  (ii) V_{os} = 1 mV, $I_b^+ = I_b^-$ = 100 nA, other effects neglected, and
  (iii) V_{os} = 1 mV, bias current compensation applied and I_{os} = 20 nA.

Note: 1 nA = 10^{-9} A.

7.2 Bias current compensation is applied to an inverting voltage amplifier circuit for which R_1 = 10 kΩ, R_2 = 250 kΩ. The unwanted component of output voltage is +50 mV. It is found that if the bias current compensation resistor is by-passed by a short circuit the unwanted output component increases to 100 mV. Calculate I_b^+.

7.3 Two identical inverting voltage amplifier circuits are connected in cascade. For each amplifier R_1 = 1 kΩ, R_2 = 5 kΩ, V_{os} = 1 mV, other op. amp. effects are negligible. Calculate the unwanted voltage component at the output of the second amplifier.

7.4 An op. amp. integrator circuit uses R_1 = 1 MΩ, C = 1 μF. The op. amp. has V_{os} = +2 mV and bias currents equal to 10 nA. How long does it take the output voltage to drift from an initial value of 0 V to a magnitude of 10 V?

7.5 Repeat Problem 7.4 with bias current compensation applied and I_{os} = 1 nA.

7.6 Estimate the bandwidth of an inverting voltage amplifier of gain −100. A 741 type op. amp. is used.

7.7 Show that for a non-inverting voltage amplifier the closed-loop gain is closely defined by $A_V' = 1 + (R_1/R_2)$ provided $|A| >> |A_V'|$. Use a Bode diagram to estimate the closed-loop bandwidth of a circuit for which A_V' = 10. The op. amp. is a 741 type.

7.8 A symmetrical square-wave signal is input to a non-inverting voltage amplifier with a gain of 10. The input signal amplitude has a peak value of 1 V and a repetition rate of f Hz. The op. amp. has a slew rate of 2 V/μs. At what value of f does the slew rate limit cause the output waveform to be completely triangular?

7.9 If signal amplitude is reduced, in general, higher signal frequencies can be used before slew rate effects occur. Referring to Worked Example 7.5, what amplitude of output sinewave must be used if slew rate effects are to place the same limit on signal frequency as does the frequency response of the op. amp. (i.e. 90 kHz)?

Selected Op. Amp. Applications 8

Objectives

- ☐ To describe some more advanced op. amp. circuits.
- ☐ To explain the operation and design of a multi-op. amp. precision difference amplifier.
- ☐ To explain the basic principles underlying the use of op. amps. in analogue computers and how the same techniques can be used for the realization of transfer functions.
- ☐ To design a Wien bridge sinewave oscillator.
- ☐ To state the inverse-function principle, and explain its general applicability.
- ☐ To describe the operation of the square wave/triangle wave circuit which contains an op. amp. *not* used in negative feedback mode.

The Chapters 6 and 7 have introduced basic op. amp. circuits and discussed the effects of the principal op. amp. non-idealities. This is by no means the whole story. Such has been the impact of op. amps. that a great many other useful op. amp. circuits have been developed. In this introductory tutorial guide only a selection can be presented. The circuits given here have been chosen because they represent a useful and interesting spread of applications and also because each contains some distinctive feature or an important principle of operation.

Precision Difference and Instrumentation Amplifiers

A circuit which can amplify the difference in two voltages has a number of useful applications. As mentioned in the section on differential amplifiers in Chapter 3, an important application is the amplification of a small signal whose source is located some distance from the amplifier. As shown in Fig. 3.11 unwanted interference appears as a common-mode voltage V_{cm} which is rejected by the difference amplifier since it has common-mode gain, A_{cm}, which is ideally zero and in practice is much lower than the differential-mode gain A_{dm}. At first glance an op. amp. operating by itself without feedback would appear to be suitable for this application. However, the wide manufacturing variations in A_{dm} (typically from -50% to $+200\%$ of the average value for A_{dm}) rule this out. A circuit offering greater precision is required and so feedback is essential.

Note that in Chapters 6 and 7, for simplicity the symbol A is used for the differential-mode gain, A_{dm}.

Another reason is that without feedback the offset V_{os} normally drives the output of the op. amp. into saturation.

The circuit shown in Fig. 8.1 is able to meet this requirement. It has two parts. One part comprises op. amp. A_3 and resistances R_1, R_2, R_3 and R_4. This is the basic precision difference amplifier. The other part, comprising resistance R_A, two matched resistances R_B, and op. amps. A_1, A_2, is an input stage which can be used to enhance the properties of the basic precision difference amplifier. The complete circuit is called an *instrumentation amplifier*.

Begin with the analysis of the basic difference amplifier. The inputs to this circuit are voltages V_1' and V_2', and the output is V_o. For simplicity the op. amp. A_3 is assumed to be ideal.

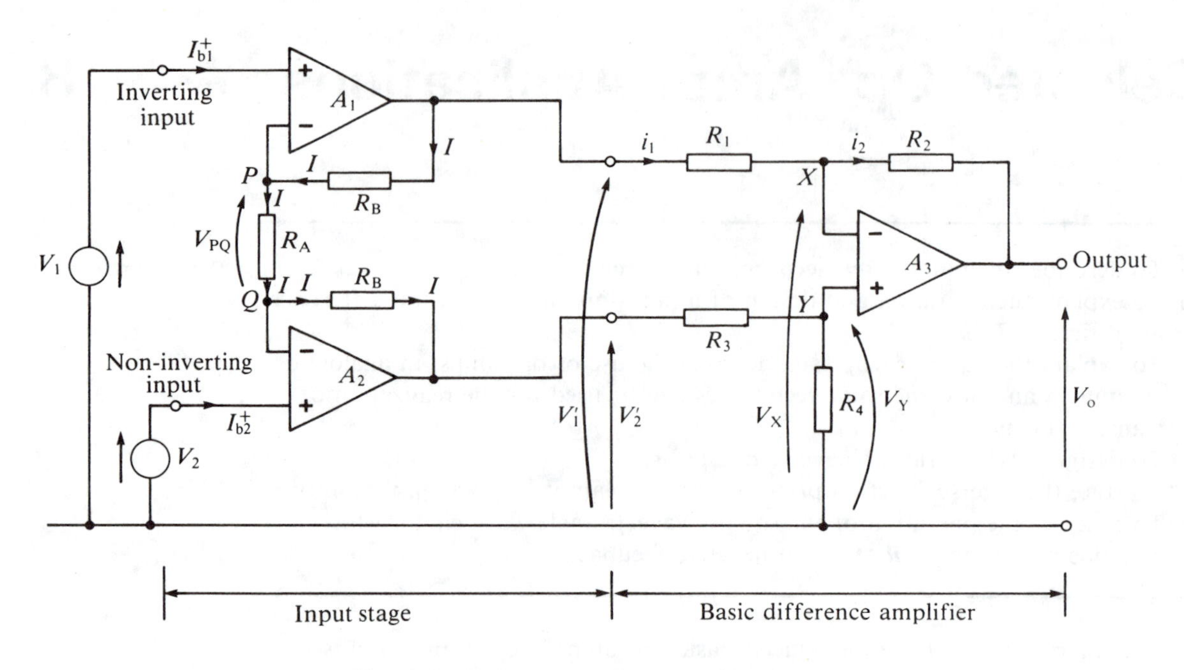

Fig. 8.1 Instrumentation amplifier.

This circuit is similar to other circuits analysed in that current i_1 enters node X from resistor R_1 and leaves through resistor R_2.

The op. amp. is assumed to draw negligible current at X.

$$i_1 = i_2 \tag{8.1}$$

Substituting for i_1 and i_2 using Ohm's law on resistors R_1 and R_2 gives

V_1' and V_O are input/output quantities which we want to keep. Variable V_X needs to be eliminated.

$$\frac{V_1' - V_X}{R_1} = \frac{V_X - V_O}{R_2} \tag{8.2}$$

Because the differential input to the op. amp. is essentially zero $V_X = V_Y$ and substituting this condition in Equation 8.2 gives

Do the manipulation yourself.

$$V_0 = (1 + \frac{R_2}{R_1})\, V_Y - \frac{R_2}{R_1}\, V_1' \tag{8.3}$$

This equation is nearly in the suitable form for a differential amplifier in that if input signals were to be applied at V_1' and V_Y both would be amplified by a factor depending on resistances R_1 and R_2, and the difference taken. The only shortcoming here is that the two multiplying factors are not equal as is required for a differential amplifier, the factor multiplying V_Y being larger than that multiplying V_1'. The reason for the potential divider formed by resistances R_3 and R_4 now becomes clear, this being to reduce the factor multiplying V_Y. For the potential divider

$$V_Y = \frac{R_4}{R_3 + R_4} \cdot V_2' = \frac{1}{1 + \dfrac{R_3}{R_4}} \cdot V_2'$$

Substituting this into Equation 8.3 gives

$$V_O = \frac{1 + \frac{R_2}{R_1}}{1 + \frac{R_3}{R_4}} \cdot V_2' - \frac{R_2}{R_1} \cdot V_1' \tag{8.4}$$

It is now possible by suitable choice of resistance values to obtain equal multiplying factors and thus provide the difference amplification of the two input signals V_2' and V_1'. Worked Example 8.1 shows how to do this.

Worked Example 8.1

Calculate suitable resistor values for the basic difference amplifier to provide a differential gain of $A_{dm} = 100$.

Solution: Equation 8.4 is required to be of the form

$$V_0 = A_{dm} V'_{dm} = A_{dm}(V_2' - V_1') = A_{dm} \cdot V_2' - A_{dm} \cdot V_1' \tag{8.5}$$

Comparing the factor multiplying V_1' in Equations 8.4 and 8.5 gives the first necessary condition: $R_2/R_1 = A_{dm}$. Comparing the factors multiplying V_2' in these expressions gives

$$A_{dm} = \frac{1 + \frac{R_2}{R_1}}{1 + \frac{R_3}{R_4}}$$

Hence,

$$A_{dm} = \frac{1 + A_{dm}}{1 + \frac{R_3}{R_4}}$$

This is easily rearranged to give the second necessary condition: $R_4/R_3 = A_{dm}$. Putting the two conditions together,

Check this yourself.

$$\frac{R_2}{R_1} = \frac{R_4}{R_3} = A_{dm} \tag{8.6}$$

Any reasonable set of resistance values which obey Equation 8.6 is usually satisfactory. For example, for $A_{dm} = 100$, one suitable set of values is $R_1 = R_3 = 1\ \text{k}\Omega$ $R_2 = R_3 = 100\ \text{k}\Omega$

The accuracy obtained for A_{dm} depends on the accuracy of the resistance values, as shown by Equation 8.6. A further point, which is sometimes more important, is that the circuit should have a low gain for common-mode signals A_{cm}. Ideally $A_{cm} = 0$. A finite A_{cm} arises because the factors multiplying V_1' and V_2' in the input-output equation (8.4) are not exactly equal. The degree of matching of these two factors depends on how well matched are the resistor ratios R_2/R_1 and R_4/R_3. To achieve a high common-mode rejection ratio very low tolerance resistors must be used (see Problem 8.4).

Common-mode rejection ratio is A_{dm}/A_{cm} (see Chapter 3).

The basic difference-amplifier can be used without the input stage with the input signals applied at the points V_1' and V_2'. A drawback of this basic circuit is that the input resistances at both input terminals are not particularly high. This can cause

Input resistances are finite because current flows into R_1 and R_3 when input signals V_1' and V_2' are applied.

loss of accuracy because of loading effects if the signal sources do not have very low self-impedances. The input stage shown in Fig. 8.1 provides the necessary buffering function and has the additional benefit of providing further gain to the circuit.

The buffering action is readily understood from an inspection of the circuit. Resistors R_B are connected in parallel feedback (voltage sensing) mode at the outputs of op. amps. A_1 and A_2 and, therefore, the output impedances are very low. These points provide the inputs V_1' and V_2' to the basic difference-amplifier. The signal inputs are now applied to the non-inverting input terminals of op. amps. A_1 and A_2. The currents at these two points in practice are very low and so the input impedances presented to the signal sources V_1 and V_2 are very high.

In practice however tiny bias currents I_{b1}^+, I_{b2}^+ flow into A_1 and A_2.

Now, consider the further amplification provided by the input stage. The input-output relationships are required, the starting point is to note that the differential inputs to A_1 and A_2 are both negligibly small. Thus point P is at essentially the same voltage as V_1, and point Q is at the same voltage as V_2. Therefore, $V_{QP} = V_1 - V_2$. The voltage V_{QP} appears across R_A and so from Ohm's law the current through the resistor is given by

$$I = \frac{V_1 - V_2}{R_A} \tag{8.7}$$

Now all of this current must flow through both resistances R_B because for ideal op. amps. no current flows into the non-inverting op. amp. terminals at modes P and Q. Therefore, the output voltage of A_1 is equal to the voltage at P (which in turn is essentially equal to V_1) plus the volt drop across upper R_B resistance associated with the current I,

$$V_1' = V_1 + IR_B \tag{8.8}$$

Using Equation 8.7 gives

$$V_1' = V_1 + (\frac{V_1 - V_2}{R_A}) \cdot R_B = V_1 (1 + \frac{R_B}{R_A}) - V_2 \frac{R_B}{R_A} \tag{8.9}$$

Repeating this argument for amplifier A_2 gives

It is not necessary to prove this from the beginning, because the circuit is up/down symmetric, Equation 8.10 can be obtained from Equation 8.9 by interchanging suffixes 1 and 2.

$$V_2' = V_2(1 + \frac{R_B}{R_A}) - V_1 \frac{R_B}{R_A} \tag{8.10}$$

It is the difference in output voltages at A_1 and A_2 which is fed to the basic difference amplifier, and from Equations 8.9 and 8.10 after collecting terms

$$(V_2' - V_1') = (1 + 2 \frac{R_B}{R_A})(V_2 - V_1) \tag{8.11}$$

It has already been shown that the differential gain of the second stage is given by the resistor ratio R_2/R_1 (also by the ratio R_4/R_3), and so the overall differential gain of the instrumentation amplifier is given by

$$A_{dm} = \frac{V_0}{V_1 - V_2} = (1 + 2 \frac{R_B}{R_A}) \frac{R_2}{R_1} \tag{8.12}$$

or equivalently

$$(1 + 2 \frac{R_B}{R_A}) \frac{R_4}{R_3} \tag{8.12}$$

Calculate suitable resistance values for the instrumentation amplifier to provide a differential gain of $A_{dm} = 100$.

Worked Example 8.2

Solution: The first task is to decide on the amplifications for the input stage and the basic difference-amplifier stage. The decision is not unique. Neither is it very critical. A reasonable sharing is for both stages to provide a differential gain of 10, giving overall $A_{dm} = 10 \times 10 = 100$. The basic difference amplifier, therefore, requires $R_2/R_1 = R_4/R_3 = 10$ and reasonable values are $R_1 = R_3 = 10\ k\Omega$ $R_2 = R_4 = 100\ k\Omega$. For the input stage Equation 8.12 is used. Substituting the known values into this equation gives

$$100 = (1 + 2\frac{R_B}{R_A}) \times 10$$

which is easily rearranged to give $R_B/R_A = 4.5$. Reasonable values for R_A, R_B which give the required ratio are $R_B = 10\ k\Omega$ and $R_B = 45\ k\Omega$

In summary, $R_1 = R_3 = R_A = 10\ k\Omega$, $R_B = 45\ k\Omega$, $R_2 = 100\ k\Omega$.

Analogue Computation

Differential equations govern the behaviour of many physical systems for, example the motion of vehicle suspensions. It can be useful to simulate part or all of a physical system by constructing an electrical analogue which has a differential equation which is the same as that of the physical system. An *analogue computer* performs this function. Although digital techniques take over much of the problem solving previously done by analogue computers, the basic idea of the analogue computer needs to be understood because it is widely used in analogue electronics. An analogue computer comprises a number of functional building-blocks which are interconnected in a particular way. The functional building blocks include op. amp. circuits which perform various functions.

As pointed out in Chapter 1 *operational* amplifiers were first invented for use in analogue computers where they are used to perform mathematical *operations* on signals.

The main functions are:

(i) *Summing* of input variables. This op. amp. circuit has been described in Chapter 6. The output signal of this type of functional block is the negative of the weighted sum of input signals. For example for three inputs V_a, V_b, V_c the output is of the form $V_0 = -(a\,V_a + b\,V_b + c\,V_c)$ where *a*, *b* and *c* are *weights* which are set by the circuit resistances. An *invertor* circuit is obtained if a single input is used and a unity weight is used.

The minus sign occurs because the summing amplifier also *inverts*.

(ii) *Summing integrator.* The basic integrator shown in Fig. 6.8 can be extended to have the following summing property for three signals by replacing resistance *R* by three resistances as shown in Fig. 6.7, $V_0 = -\int(a\,V_a + b\,V_b + c\,V_c)\cdot dt + V_0(0)$. This technique can be extended to any number of inputs.

Again, the integrator circuit *inverts* as well as integrates; hence the minus sign.

(iii) *Non-linear functional blocks.* These include electronic circuits to perform the multiplication of two signals and also the generation of non-linear functions.

A detailed treatment of analogue computer techniques including non-linear functional blocks can be found in specialist texts, for example, Hyndman, D.E. *Analog and Hybrid Computing* (Pergamon, 1970).

The basic idea of an analogue computer is to interconnect op. amp. circuits in such a way that some of the nodal voltages are forced to become analogues of physical variables in the system under investigation. To show this consider the solution of a second-order linear differential equation. In its simple form this is,

$$\frac{d^2y}{dt^2} + a_1 \frac{dy}{dt} + a_0 y = b_0 \cdot x(t) \quad (8.13)$$

Here y is the unknown and $x(t)$ is the input *forcing function*. In the most general form of second-order differential equation derivatives of the forcing function up to second degree can appear on the right-hand side. Thus

$$\frac{d^2y}{dt^2} + a_1 \frac{dy}{dt} + a_0 y = b_2 \frac{d^2x}{dt^2} + b_1 \frac{dx}{dt} + b_0 x \quad (8.14)$$

The differential terms in this equation cannot in practice be directly simulated by differentiator op. amp. blocks for the reason given in Chapter 6. The differential equation has to be transformed into an equation involving integral operations and which has exactly the same solution. Then the required circuit arrangement can easily be drawn as shown, for example, in Fig. 8.2a and b. The derivation of these circuits for the differential Equation 8.14 is as follows. The method used is called the method of *successive integration*. It is a general method which is applicable to differential equations of any degree. The basic strategy is to keep terms involving differentials on the left-hand side and then successively integrate until all differentials have been removed. The resulting form of the equation is then directly realisable as an analogue computer. In what follows the known coefficients a_1, a_2 and b_0, b_1, b_2 are assumed to be positive.

The first step is to arrange the differential equation so that all differential terms are on the left and all non-differential terms on the right. Equation 8.14 becomes

$$\left(\frac{d^2y}{dt^2} - b_2 \frac{d^2x}{dt^2}\right) + \left(a_1 \frac{dy}{dt} - b_1 \frac{dx}{dt}\right) = (b_0 x - a_0 y) \quad (8.15)$$

The next step is to integrate both sides and also multiply throughout by -1. This gives

Remember *integration* is the reverse operation to *differentiation*.

$$-\left(\frac{dy}{dt} - b_2 \frac{dx}{dt}\right) - (a_1 y - b_1 x) = -\int (b_0 x - a_0 y) \cdot dt \quad (8.16)$$

Defining $f_1 = -\int (b_0 x - a_0 y) dt$, Equation 8.16 is written as (8.17)

$$-\left(\frac{dy}{dt} - b_2 \frac{dx}{dt}\right) - (a_1 y - b_1 x) = f_1 \quad (8.18)$$

Comparing this with Equation 8.15 it can be seen that the order of differentiation on the left-hand side has been reduced by one. Now repeat this process on Equation 8.18. Arranging non-differential terms on the right gives

$$-\left(\frac{dy}{dt} - b_2 \frac{dx}{dt}\right) = (a_1 y + f_1 - b_1 x)$$

Integrating both sides and multiplying by -1 gives

$$(y - b_2 x) = -\int (a_1 y + f_1 - b_1 x) dt \quad (8.19)$$

Defining $f_2 = -\int (a_1 y + f_1 - b_1 x) dt$ gives (8.20)

$$y - b_2 x = f_2 \quad (8.21)$$

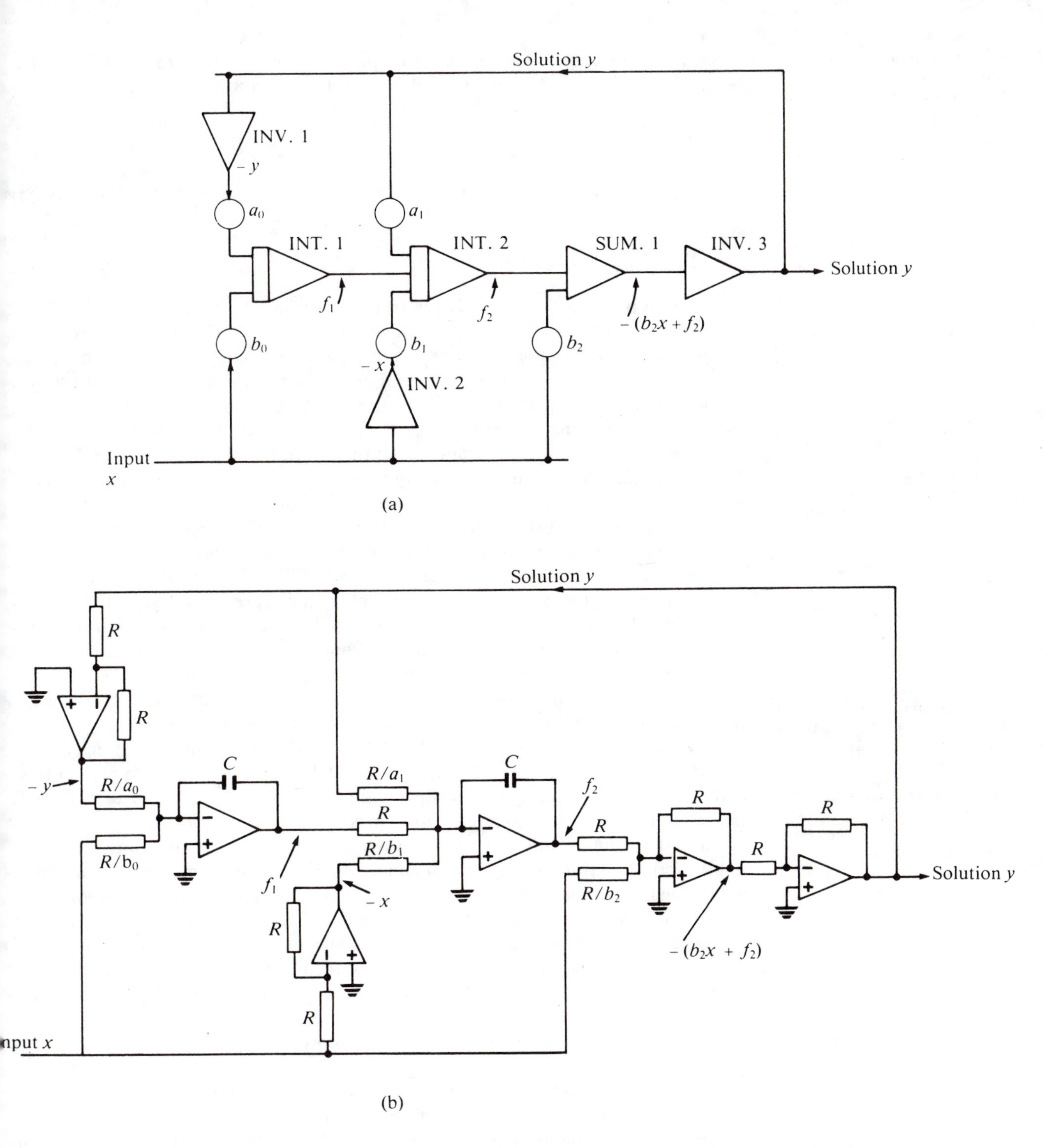

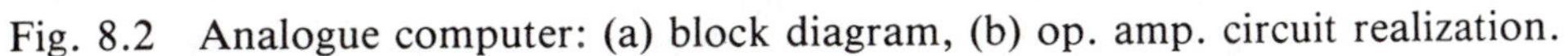

Fig. 8.2 Analogue computer: (a) block diagram, (b) op. amp. circuit realization.

For a third-order or higher order differential equation, the process would have continued some more steps.

All differential terms have now been removed. The solution y to the original differential Equation 8.14 is also the solution to Equation 8.21. That is,

$$y = b_2 x + f_2,$$

where

$$f_2 = -\int(a_1 y + f_1 - b_1 x)\mathrm{d}t \qquad (8.22)$$

and

$$f_1 = -\int(b_0 x - a_0 y)\mathrm{d}t$$

This group of conditions (8.22) is in a form which is directly realizable as an analogue computer. The result is shown in Fig. 8.2a. The symbols are those commonly used for analogue computer representations and their meanings are made clear later. Since this system diagram will be realised as the electrical system shown in Fig. 8.2b, the input x and the solution y are to be voltages and the functional blocks are operational amplifier circuits. The realization commences by drawing lines representing the input variable x and the solution variable y in Fig. 8.2a. Next Equations 8.22 are implemented in sequence starting with f_1, then f_2 and finally y. Referring to the diagram the point f_1 is seen to be the output of integrator INT 1. Chapter 3 showed that the op. amp. integrator is able to provide the negative of the integral of the sum of inputs. INT 1 symbol performs this function, $f_1 = -\int(b_0 x - a_0 y)\cdot\mathrm{d}t$ (see Equations 8.22). The two inputs to INT 1 come from y via the coefficient multiplier a_0, (coefficient multipliers are represented by circle-symbols) and from x via b_0. The invertor INV 1 is inserted in the a_0y path because f_1 requires signal $-a_0y$ at the input to the integrator. Function f_2 in Equations 8.22 is realized by integrator INT 2. As the defining equations indicate, the inputs to INT 2 are a_1y, f_1, and $-b_1x$. The invertor INV 2 is included in the b_1x path to introduce the required minus sign. Finally Equations 8.22 show the solution y is given by the sum of b_2x and f_2. The building block SUM 1 sums the signals b_2x and f_2, but in doing so introduces a minus sign, so the output of SUM 1 is $-(b_2 x + f_2)$. To correct this, and provide the solution to the differential equation y, the invertor INV 3 is included. This completes the realization process.

The reasons for multiplying by −1 as well as integrating in the successive approximation method is now clear; it allows the use of the inverting op. amp. integrator circuit.

SUM 1 can be made from the basic inverting voltage amplifier using the summing property for several inputs. See Chapter 6.

Negative differential equation coefficients (a_1, a_2, b_2, b_1 or b_0) can be accommodated by the use of further invertors.

On first encounter this procedure for realizing the analogue computer may seem dubious; the solution line y is drawn and apparently y is presumed to exist before it is obtained from solution equation $y = b_2 x + f_2$ as implemented in SUM 1 and INV 3. However, there is no fallacy here. This is because, although the sequence is followed in realizing the analogue computer, when it is switched on, the input signal x, flows first through the integrators to y and then back. When the system has been energized the voltages at the various nodes must satisfy the above equations. Thus, the desired solution $y(t)$ is generated automatically.

A commercial analogue computer often comprises a selection of building blocks which are connected together by the user in accordance with the realization procedure. The alternative way is to wire up the circuits individually to make a special purpose analogue computer if only one particular system is to be simulated. Using the basic circuits of Chapter 6, Fig. 8.2a can be constructed as shown in Fig. 8.2b. Any convenient values can be chosen for R and C for the individual op. amp. circuits; it is assumed that for the integrators $RC = 1$ s, (e.g. $C = 1\ \mu$F, $R = 1$ MΩ).

Before leaving this topic the question of *transfer function* realization is touched upon. It can be seen from Chapter 6 that a differentiator circuit can be looked at in two ways, first in the time domain

$$v_{out} = -CR \frac{d\,v_{in}}{dt}$$

or second when the signals are sinusoidal with angular velocity ω,

$$\hat{V}_{out} = (-CR)\,j\omega \cdot \hat{V}_{in}$$

In other words, for steady-state sinusoidal equations the differential operator d/dt can be replaced by $j\omega$. Applying this to the original differential equation (Equations 8.14) it may be written as

$$(j\omega)^2\hat{Y} + a_1(j\omega)\hat{Y} + a_0\hat{Y} = b_2(j\omega)^2\hat{X} + b_1(j\omega)\hat{X} + b_0\hat{X}$$

Readers familiar with Laplace transform techniques will know the more general use of the complex frequency variable $s = \sigma + j\,\omega$. The transfer function is

$$\frac{b_0 + b_1 s + b_2 s^2}{1 + a_1 s + a_2 s^2}$$

That is

$$\frac{\hat{Y}}{\hat{X}} = \frac{\text{Output}}{\text{Input}} = \frac{b_0 + b_1(j\omega) + b_2(j\omega)^2}{1 + a_1(j\omega) + a_2(j\omega)^2} \qquad (8.23)$$

This formulation shows that the analogue computer in Fig. 8.2a can be looked at as a realization of the general second order transfer function (Equation 8.23). This transfer function is in the form of the ratio of two polynomials in ω. Hence the behaviour of $\hat{Y}/\hat{X}$ is a function of signal frequency. A circuit which realizes a desired transfer function is called a *filter*. Filters play an important part in many subject areas such as communications, and automatic control. Methods are known to choose the coefficients of a transfer function to achieve some desired filtering behaviour. Thus it can be seen that analogue computer techniques can be used as a method for realizing such transfer functions.

The analogue computer shown in Fig. 8.2b is made from resistors, capacitors and op. amps. Op. amps. are active devices and so if a realisation of the type in Fig. 8.2b is used as a filter it is called an *RC-active* filter. Active filters are very important and many techniques for their realization are known.

Active filters are dealt with in more detail in the next chapter.

Wien Bridge Oscillator

A sinewave oscillator is a source of a continuous sinusoidal signal. Oscillators have many uses, just one of them being to provide a test signal for the measurement of amplifier performance. The main features desired of a sinewave oscillator are well-defined frequency and amplitude, and also low distortion in the waveform.

As with any signal source the circuit has an output but no input.

For high frequencies *LC* oscillator circuits are commonly used, in which the output signal frequency is defined by one or more inductors and capacitors. At lower frequencies larger and more costly inductors are required in an *LC* oscillator, and *RC oscillators* are usually favoured. *RC* oscillators use resistors and capacitors to define the frequency.

A very popular *RC* oscillator is the *Wien bridge* oscillator, shown in Fig. 8.3a. It comprises two main sections, a non-inverting voltage amplifier and a frequency selective *RC* circuit called a *Wien network*. The output of the voltage amplifier is fed-back to the input of the frequency selective circuit. Thus we have a feedback circuit with the frequency selective circuit providing the voltage feedback fraction β_V.

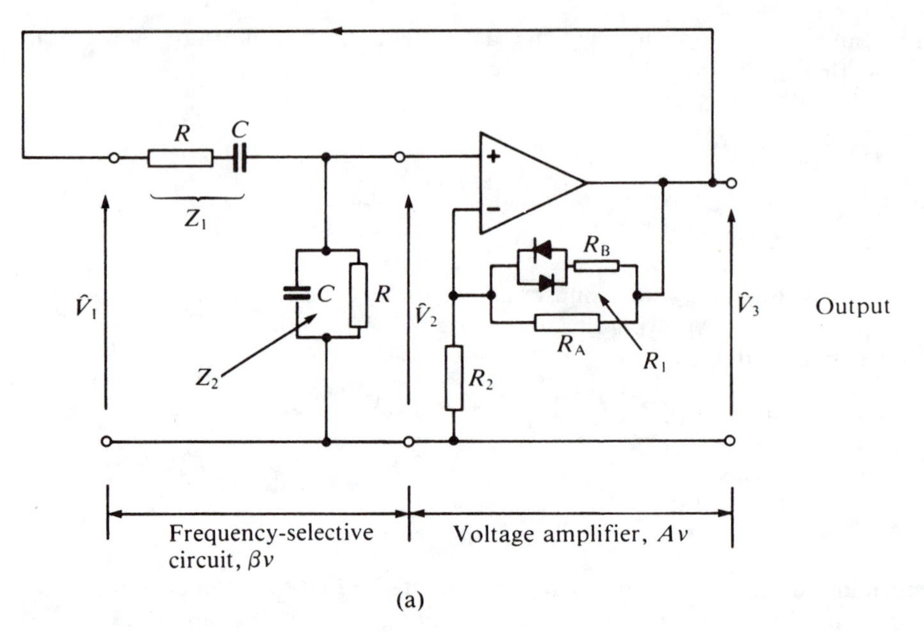

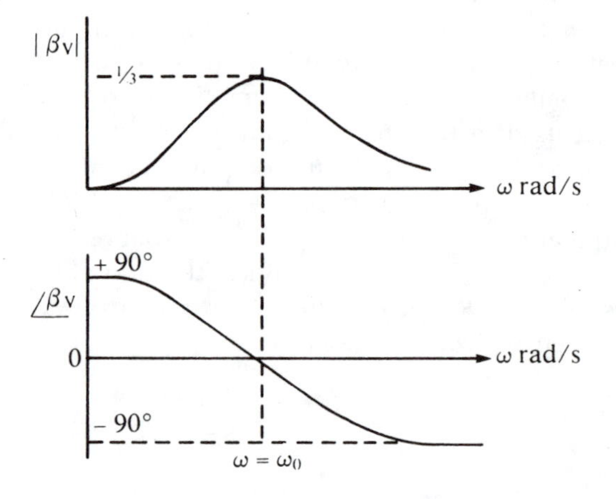

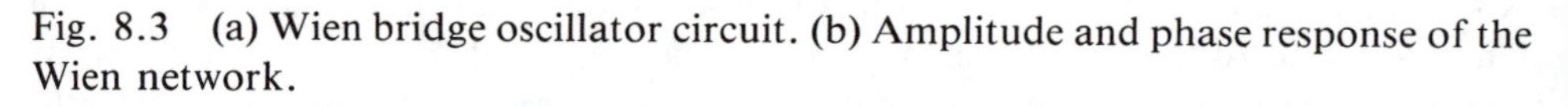

Fig. 8.3 (a) Wien bridge oscillator circuit. (b) Amplitude and phase response of the Wien network.

To understand how the circuit works first look at the frequency selective circuit. Basically it is a simple potential divider formed by impedances Z_1 and Z_2. Hence

This is recognised as the voltage coupling factor for impedances Z_1 and Z_2.

$$\frac{\hat{V}_2}{\hat{V}_1} = \beta_V = \frac{Z_2}{Z_1 + Z_2} = \frac{1}{1 + Z_1/Z_2} \tag{8.24}$$

The impedance Z_1 is the series combination of a capacitance and resistance and therefore in complex impedance form is given by

$$Z_1 = R + \frac{1}{j\omega C}$$

The impedance Z_2 is the parallel combination of a capacitor and resistor, and therefore the admittance $1/Z_2$, of the combination is given by

$$\frac{1}{Z_2} = \frac{1}{R} + j\omega C$$

Substituting these results into Equation 8.24 gives

$$\beta_V = \frac{1}{1 + (R + \frac{1}{j\omega C})(\frac{1}{R} + j\omega C)}$$

After expanding the denominator and collecting real and imaginary terms, this becomes

$$\beta_V = \frac{1}{3 + j(\omega CR - \frac{1}{\omega CR})} \tag{8.25}$$

The amplitude and phase of β_V vary with frequency as shown in Fig. 8.3b. The peak in $|\beta_V|$ occurs at an angular frequency ω_0 which causes the frequency-dependent imaginary term in the denominator to be zero. That is,

Confirm this expression and satisfy yourself that the frequency-response curves are of the form shown.

$$\omega_0 CR - \frac{1}{\omega_0 CR} = 0$$

which provides the result $\omega_0 = 1/CR$. The phase angle $\angle\beta_V$ is zero since the imaginary term is zero.

Thus at ω_0

$$\beta_{V_0} = \frac{1}{3 + j(\text{zero})} = \frac{1}{3} \quad \text{and} \quad \angle\beta_{V_0} = 0 \tag{8.26}$$

The task of the amplifier is to sustain oscillations by supplying sinusoidal energy to the circuit and to any load connected at the output. Since the non-inverting amplifier has zero phase shift, to complete the loop the Wien network must also be operating at zero phase shift and therefore at the angular frequency ω_0. If the amplifier has a gain of three, $A_V = 3$, this balances the attenuation of the Wien network and oscillations are self sustained.

For oscillations to start $A_V > 3$ is required. When the circuit is switched on, small random perturbations which are always present in a circuit pass round the loop and the frequency component at $\omega = \omega_0$ builds up. This raises the question of *amplitude control*. When the circuit is switched on $A_V > 3$ is required but at the required amplitude the condition $A_V = 3$ must be exactly maintained. When the circuit is running if the sinewave amplitude starts to fall or rise the amplifier gain must be raised or lowered to maintain the oscillation amplitudes. Fig. 8.3a shows one possible amplitude control arrangement. The usual resistor R_1 in the non-inverting amplifier is replaced by a diode-resistor arrangement which acts as an amplitude sensitive resistor. Its action is made clear in the following worked example.

Worked Example 8.3

Design a Wien bridge oscillator to oscillate at a frequency of 10 kHz.

Solution: The frequency of oscillations is defined by the frequency-selective circuit for which $\omega_0 = 1/CR$. Turning this equation around,

$$CR = \frac{1}{\omega_0} = \frac{1}{2\pi f_0} = \frac{1}{2\pi \times 10\ k} = 1.59 \times 10^{-5}\ s \qquad (8.27)$$

A wide range of values for C and R can satisfy this condition. If C is very small (say less than 100 pF) then the actual C obtained is affected by stray capacitances. If C is very large (say greater than 1 μF). The cost of good quality capacitors becomes high. A good compromise is $C = 0.01\ \mu$F. With C decided, R can be calculated to satisfy Equation 8.27,

$$R = \frac{1}{\omega_0 C} = \frac{1.592 \times 10^{-5}}{C} = \frac{1.592 \times 10^{-5}}{0.01 \times 10^{-6}} = 1.59\ k\Omega$$

Hence $C = 0.01\ \mu$F, $R = 1.59$ kΩ provides a frequency of 10 kHz. The non-inverting amplifier needs to have a gain of 3 at the maintained oscillation amplitude. Thus

This is the gain expression for the basic non-inverting voltage amplifier (see Equation 6.31).

$$A_V = (1 + \frac{R_1}{R_2}) = 3$$

Again the choice for R_1 and R_2 is not unique. One of many suitable values for R_2 is 10 kΩ. This means $R_1 = 20$ kΩ is required.

The amplitude control mechanism shown in Fig. 8.3b must now be considered. Since the amplitude grows if $A_V > 3$ and decreases if $A_V < 3$, A_V must be arranged to vary automatically from, say, 3.2 when the amplitude is too low to, say, 2.8 when the amplitude is too high. The two diodes shown in Fig. 8.3a make this possible.

On each half cycle one diode is forward biased and owing to the exponential characteristic its effective resistance decreases as the amplitude of the signal builds up. This reduces the effective value of R_1 and hence the gain of the amplifier. $A_V > 3$ is required to cause oscillation amplitude to build up and so R_1 greater than 20 kΩ is required. It is sufficient to choose $R_A = 22$ kΩ to give

$$A_V = (1 + \frac{22\ k}{10\ k}) = 3.2$$

The effective resistance of the two diodes decreases as the amplitude rises until eventually R_B is effectively placed in parallel with R_A thus giving

$$R_1 \approx \frac{R_A R_B}{R_A + R_B}$$

which is lower than the previous value $R_1 \approx R_A$ and thus reduces the amplifier gain below $A_V = 3$. For $A_V = 2.8$, $R_1 = 18$ kΩ is required to give $[1 + (R_1/R_2)] = 2.8$. The final step is to calculate a value for R_A which when combined in parallel with $R_B = 22$ kΩ gives the required value of $R_1 = 10$ kΩ. This value is 220 kΩ.

The relationship used here is

$$R_B = \frac{R_1 R_A}{R_A - R_1}$$

Prove this yourself.
Negative temperature coefficient *thermistors* are also used for R_1. For larger voltages they heat up and have reduced resistance, thus reducing the voltage gain.

In operation the circuit amplitude stabilises itself so that the diodes partially conduct to give an effective resistance $R_1 = 20$ kΩ and so ensure $A_V = 3$. The calculation of the output amplitude at which this occurs is not easy and is beyond the scope of this text.

In summary $R = 1.59$ kΩ, $C = 0.01\ \mu$F, $R_A = 22$ kΩ, $R_B = 220$ kΩ.

The Inverse Function Principle

The *inverse function* principle can be understood by considering the usual feedback

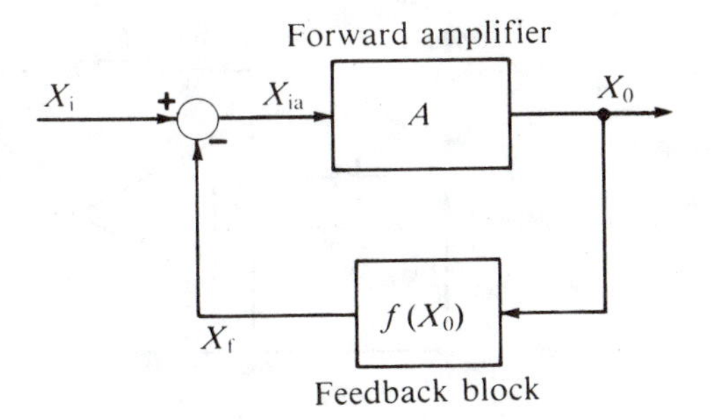

Fig. 8.4 Feedback block diagram showing non-linear feedback block.

block diagram shown in Fig. 8.4. This block diagram differs from others discussed in that the feedback block is now considered to be some known *non-linear* function, $X_f = f(X_0)$.

At the summing junction

$$X_{ia} = X_i - X_f = X_i - f(X_0) \tag{8.28}$$

Suppose now the forward amplifier has very high gain so that $X_{ia} = X_0/A \approx 0$. Then Equation 8.28 becomes

$$0 = X_i - f(X_0)$$

Hence

$$f(X_0) = X_i$$

Finally taking the inverse function of both sides, gives

$$X_0 = f^{-1}(X_i) \tag{8.29}$$

This established the inverse function principle.

This shows that the feedback system provides an output which is the inverse function of the input. Before looking at real applications of the principle it is interesting to consider what happens if a *linear* feedback block is assumed, $X_f = \beta \cdot X_0$. The inverse function principle, Equation 8.29, shows that provided A is very high $X_0 \approx X_i/\beta$. This result is already familiar from Chapter 2.

Now two applications of the principle to non-linear op. amp. circuits are discussed. Conveniently an op. amp. can provide the high forward gain required.

The first application uses the basic exponential current-voltage relationship of a semiconductor diode,

$$I_D = I_0(\exp \cdot \frac{V_D}{V_0} - 1) \tag{8.30}$$

where I_D, V_D are the diode current and voltage, I_0 is the diode saturation current, typically 1 μA, and V_0 is a constant typically 25 mV, known as the thermal voltage.

In forward bias for all except tiny diode voltages, the exponential term is much greater than -1 and so Equation 8.30 can be written

$$I_D \approx I_0 \exp \cdot \frac{V_D}{V_0} \tag{8.31}$$

The exponential diode equation.

Fig. 8.5a shows the diode incorporated in the feedback path of an inverting voltage amplifier, and because of the inverse function principle the circuit provides

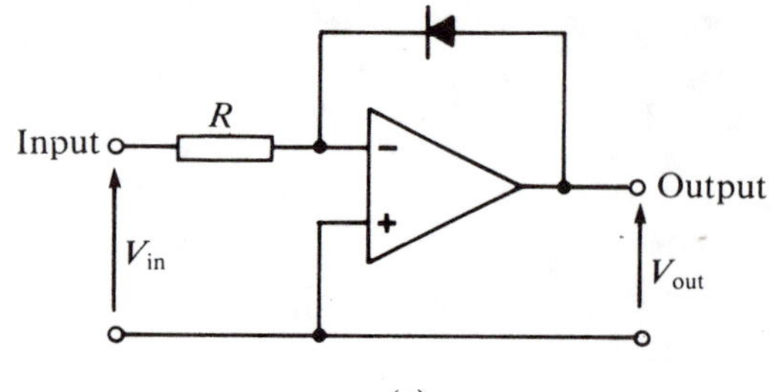

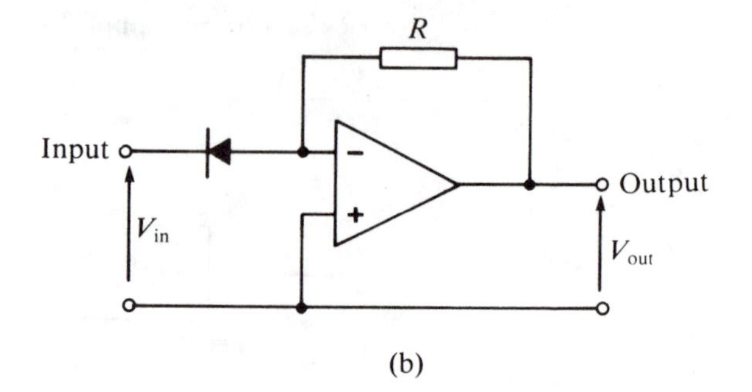

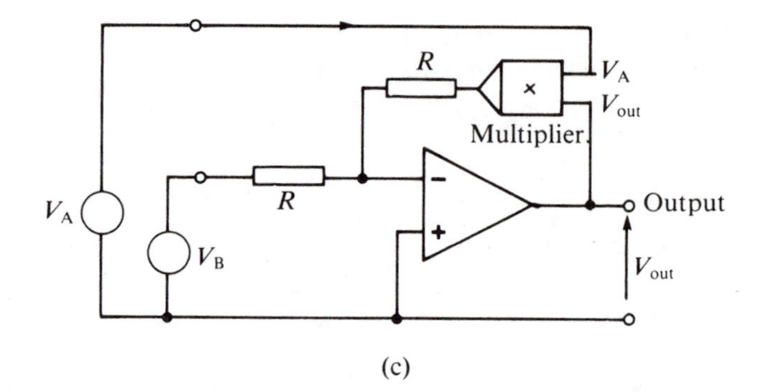

Fig. 8.5 Applications of the inverse-function principle: (a) log circuit, (b) antilog circuit, (c) divider.

log is the inverse of the exponential function. The exponential function is the inverse of the logarithm function; that is exp ≡ antilog.

an output which is proportional to the logarithm of the input voltage. The companion circuit Fig. 8.5b provides the basic exponential (that is the antilogarithm) function. Log and antilog circuits are quite useful (see Problem 8.7).

The second example uses an analogue multiplier. Analogue multipliers produce a voltage which is proportional to the product of two input voltages. They are available in integrated-circuit form at relatively low cost. The inverse of multiplication is division and Fig. 8.5c shows the circuit to provide an output voltage which is proportional to one voltage divided by another.

This circuit has two inputs.

Exercise 8.1 Assuming ideal op. amps prove that the input-output relationships of the circuits in Fig. 8.5 are given by

(a) $V_{out} \approx V_0 \log_e\left(-\dfrac{V_{in}}{I_0 R}\right)$

(b) $V_{out} \approx -I_0 R \exp \cdot \dfrac{V_{in}}{V_0}$

(c) $V_{out} = \left(-\dfrac{1}{K}\right) \cdot \dfrac{V_B}{V_A}$, assuming the output of the multiplier is $K\,V_A\,V_{out}$.

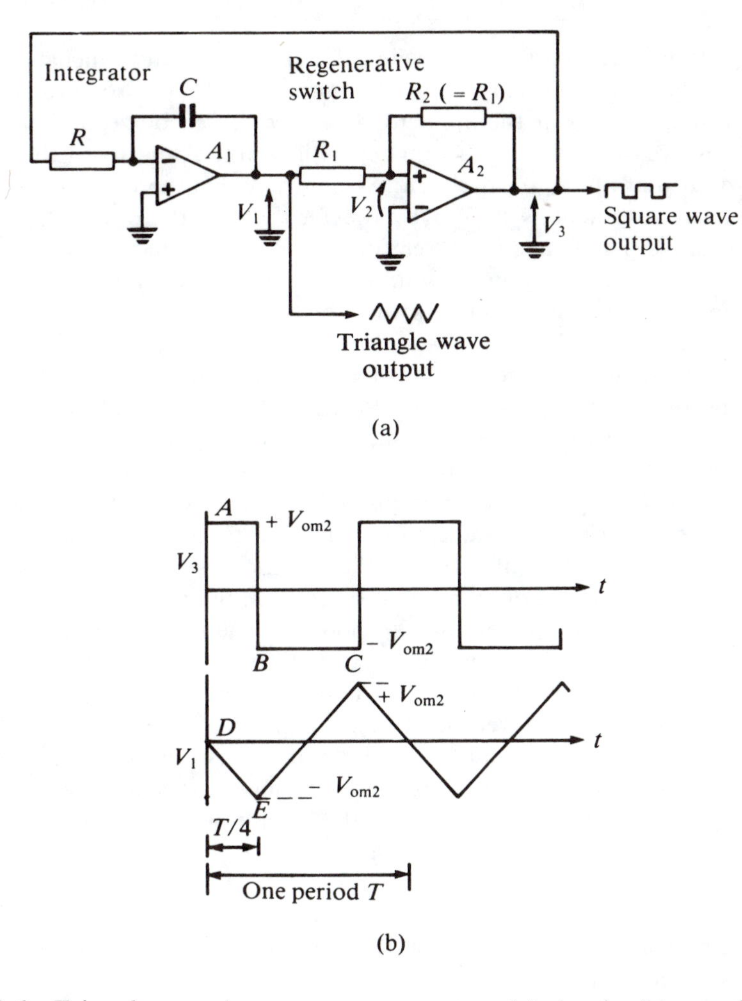

Fig. 8.6 Triangle wave/square wave generator: (a) circuit, (b) waveforms.

Triangle Wave/Square Wave Generator

This circuit (see Fig. 8.6a) generates two output waveforms as shown in Fig. 8.6b. One waveform is a repetitive triangle at V_1, and the other is a repetitive square wave at V_3.

Again because the circuit is a generator there is no input signal. There are however two outputs in this case.

The square wave at V_3 is fed back to the integrator which converts it to a triangular waveform. This waveform is passed to the regenerative-switch which converts it to a square wave, thus providing a self-sustained operation.

It is important to note that the circuit comprising R_1, R_2, A_2 is *not* an inverting voltage amplifier as might at first seem. In this case feedback resistor R_2 is connected to the *non-inverting* input terminal of the op. amp. Therefore, the feedback is positive. Because the op. amp. gain is high the amount of positive feedback is also high. The amplifier is driven into saturation and the input is *not* constrained by the virtual earth principle. To simplify the discussion assume $R_1 = R_2$. Because these resistors are equal and negligible current flows into the op. amp. non-inverting input terminal the voltage V_2 always lies midway between voltage V_1 and V_3.

For the analysis where $R_1 \neq R_2$ see Clayton, G.B. *Operational Amplifiers*, 2nd ed. (Newnes-Butterworth, 1979) pp. 279–288.

When the circuit is first switched on the capacitor is uncharged and so $V_1 = 0$. At the instant of switch on V_3, the output of A_2, is also zero. However, tiny offset voltage effects at the input to A_2 are amplified by A_2 and cause V_3 to move in a positive or negative direction depending on the polarity of the offset effects for the particular op. amp. specimen. Assume here that the V_3 moves in a positive direction. This movement is passed by the feedback path to V_2. Because the feedback is positive the change in V_2 reinforces the original movement in V_3, which in turn is passed back to V_2 and so on. Very quickly this regenerative action drives the op. amp. output into positive saturation $+V_{om2}$. This corresponds to point A on the waveform in Fig. 8.6b.

The saturation voltage V_{om} is typically 1 V less than the op. amp. power supply voltage.

The steady voltage $+V_{om2}$ is fed to the integrator circuit which causes V_1 to move negatively at a constant rate determined by R and C. The voltage V_2 which lies midway between $V_3 = V_{om2}$ and V_1 is pulled down by the falling movement of V_1. Eventually when V_1 has fallen sufficiently to have a value equal and opposite to V_3 (that is $V_1 = -V_{om2}$) the mid-point V_2 is at zero volts. This causes the op. amp. A_2 to come out of positive saturation and V_3 starts to move negatively. This movement is fed to V_2 which reinforces the negative movement in V_3. Again regenerative action quickly forces the op. amp. into saturation, but this time into negative saturation $V_3 = -V_{om2}$. This corresponds to point B. The integrator now has a steady negative input voltage $-V_{om2}$ and so V_1 starts to move in a positive direction at a constant rate. This carries on until V_1 is equal and opposite to V_3 (that is $V_1 = +V_{om2}$) when V_2 crosses zero and regenerative switching in A_2 occurs (point C). The voltage V_3 is now positive again and so the cycle of events repeats itself, and does so over and over again to produce the repetitive triangle and square waveforms.

The reader may notice that the transition at point B only occurs if the magnitude of the saturation voltage V_{om1} of A_1 is not less than that the saturation voltage V_{om2} of A_2, since otherwise V_1 could not reach the required value $-V_{om2}$ needed to trigger A_2. This difficulty can be avoided by making $R_1 < R_2$, but the analysis is then more complex and is left to the reader.

An important parameter of the circuit is the frequency of oscillation. The following worked example deals with this.

Worked Example 8.4 Derive an expression for the frequency of oscillation of the triangle wave/square wave generator of Fig. 8.6.

Solution: The time for one cycle, T, is marked on the waveform in Fig. 8.6b. From the symmetry of the waveform it can be seen that the time it takes for the triangle waveform to move from point D to E is equal to a quarter of a cycle, $T/4$. In this section of the waveform the integrator is integrating an input voltage $V_1 = +V_{om2}$ and point E occurs when the integrator output equals $-V_{om2}$. Hence, the general equation for the integrator

$$v_{out} = -\frac{1}{RC}\int v_{in}\,dt$$

can be used to evaluate V_1 after a quarter of a period as follows

$$-V_{om2} = V_1 = -\frac{1}{RC}\int_0^{T/4}(+V_{om2})\cdot dt$$

Definite integration and then substitution of limits.

$$= -\frac{1}{RC}\Big[V_{om2}\cdot t\Big]_0^{T/4} = \frac{-V_{om2}}{RC}\cdot\frac{T}{4} \qquad (8.32)$$

Therefore $T = 4\,RC$ and the frequency of oscillation is $1/4\,RC$.

Summary

Some applications of op. amps. have been presented. One of these is the single op. amp. precision difference amplifier. Provided the four resistors are accurately matched to satisfy $R_2/R_1 = R_4/R_3$, the differential gain is given by $A_{dm} = R_2/R_1$ and the common mode gain A_{cm} is zero. By adding a two op. amp. input stage this circuit provides very high input impedance to both signal inputs and also further amplification. This whole three op. amp. circuit is called an instrumentation amplifier.

The second application was the use of op. amps. to make an analogue computer. Analogue computers are useful for simulating differential equations. The realization of a general second-order differential equation has been presented using the method of successive integration. This method can be applied to differential equations of any order. Analogue computers provide one way of achieving transfer function realization, which introduces the important topic of active filters.

If a circuit or component having a non-linear characteristic is used in the negative feedback path of a high-gain amplifier the inverse function is generated by the circuit. This is the inverse function principle. It shows how feedback principles can be used to invert the function to provide an input/output relationship of the form out $= f^{-1}$(in). Two examples of the use of this principle are the logarithmic amplifier and the use of a multiplier to make a divider circuit.

The final application described was the triangle wave/square wave generator. The triangle wave and square wave outputs of the circuit described have equal amplitudes determined by the saturation voltage of one of the op. amps. The circuit differs from the others in that one op. amp. is used in a regenerative switching mode. In this mode positive feedback is present and the virtual earth principle does not apply.

Other op. amp. circuits are to be found in the next chapter and in the books listed in the bibliography.

Problems

8.1 If all resistors are equal in the instrumentation amplifier, what is A_{dm}?

8.2 A_{dm} of the instrumentation amplifier can be varied by a single resistor. Which resistor is it?

8.3 Design an instrumentation amplifier to give $A_{dm} = 400$. Choose $R_A = R_1 = 5\ \text{k}\Omega$ and design the input stage to have the same differential-mode gain as the basic difference amplifier stage.

8.4 A basic difference amplifier is designed to have resistance values $R_1 = R_3 = 1\ \text{k}\Omega$ and $R_2 = R_4 = 100\ \text{k}\Omega$. When constructed all resistances are accurate except R_3 which is too high in value by 1%. Assume the op. amp. is ideal. Calculate the differential-mode gain, common-mode gain, and common-mode rejection ratio of the circuit.

8.5 Suppose the general second-order differential equation is multiplied throughout by (-1):

$$-\frac{d^2y}{dt^2} - a_1\frac{dy}{dt} - a_0 y = -b_2\frac{d^2x}{dt^2} - b_1\frac{dx}{dt} - b_0 \qquad (8.36)$$

The solution y must be unaffected by this. Use the successive integration method on Equation 8.36 to provide an alternative analogue computer realization to Fig. 8.2b.

8.6 Design a Wien bridge oscillator to oscillate at 2 kHz. Assume $C = 0.01\ \mu\text{F}$.

8.7 Suppose the resistors and capacitors in the Wien bridge oscillator are made unequal: $Z_1 = (R_1$ in series with $C_1)$, $Z_2 = (R_2$ in parallel with $C_2)$. Re-analyse the circuit and show that $\omega_0 = 1/(R_1R_2C_1C_2)$ provided

$$A_V \geqslant 1 + \frac{R_1}{R_2} + \frac{C_2}{C_1}.$$

8.8 Use the log/antilog circuits together with any other basic op. amp. circuits required, to create circuits which provide the following functions

(i) $V_{out} \propto V_{in1} \times V_{in2}$
(ii) $V_{out} \propto V_{in}^2$
(iii) $V_{out} \propto \sqrt{V_{in}}$

Further Op. Amp. Applications 9

Objectives

- ☐ To describe more op. amp. circuits.
- ☐ By taking the precision difference amplifier as a starting point, to show how interesting circuits for other applications can be generated.
- ☐ To explain the purpose and operation of a.c. amplifiers.
- ☐ To show how to design active filters.
- ☐ To explain the precision half-wave rectifier circuit and how it can be used to make a precision full-wave rectifier and a.c./d.c. converter.

The previous chapter presents op. amp. circuits which are selected because of distinctive features or important principles of operation. This chapter extends this by widening out the variety and range of op. amp. circuit applications.

Circuits Derived from the Precision Difference Amplifier

The op. amp. circuits to be described here can all be derived from the single op. amp. precision difference amplifier presented in the previous chapter. They were not necessarily originally discovered in this way (although they might have been) but in presenting them this way they serve to illustrate that new applications can often be found by looking at already known circuits in the right way. The circuits also provide useful applications.

The basic difference circuit appears in Fig. 8.1 of the previous chapter. It has two inputs, V_1' and V_2'. For the present purpose these input points are joined together and a single input signal V_i is applied. The circuit now appears as shown in Fig. 9.1.

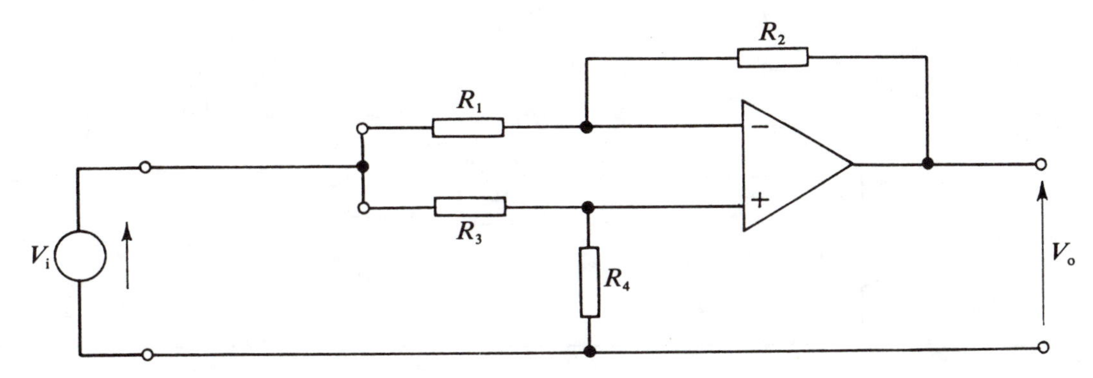

Fig. 9.1 Difference amplifier with joined inputs.

To obtain the input–output relationship for this circuit, there is no need to perform an analysis from the very beginning. The input–output relationship has already been obtained for separate inputs V_1' and V_2' and appears as Equation 8.4. All that is required is to replace V_1' and V_2' in that equation by V_i. The following is obtained:

Check this yourself.

$$V_o = \frac{1 + \dfrac{R_2}{R_1}}{1 + \dfrac{R_3}{R_4}} \cdot V_i - \frac{R_2}{R_1} \cdot V_i$$

which gives the voltage transfer function

Confirm this yourself.

$$A_v = \frac{V_o}{V_i} = \frac{1 - \dfrac{R_2}{R_1} \cdot \dfrac{R_3}{R_4}}{1 + \dfrac{R_3}{R_4}} \tag{9.1}$$

Now for the derived circuits: the switched gain-polarity circuit; the variable bi-polar gain circuit; the bridge amplifier; and the phase shifter. They are all obtained by constraining or replacing resistances R_1, R_2, R_3 and R_4 in different ways.

The switched gain-polarity circuit

The circuit is shown in Fig. 9.2. Resistors R_1 and R_2 are set equal to some value, R. Resistor R_4 is replaced by a switch S. When the switch is open this is equivalent to having R_4 equal to infinity. Inserting these conditions into the voltage transfer function (9.1),

$$A_v = \frac{V_o}{V_i} = \frac{1 - \dfrac{R}{R} \cdot \dfrac{R_3}{\infty}}{1 + \dfrac{R_3}{\infty}} = \frac{1 - 0}{1 + 0} = +1 \tag{9.2}$$

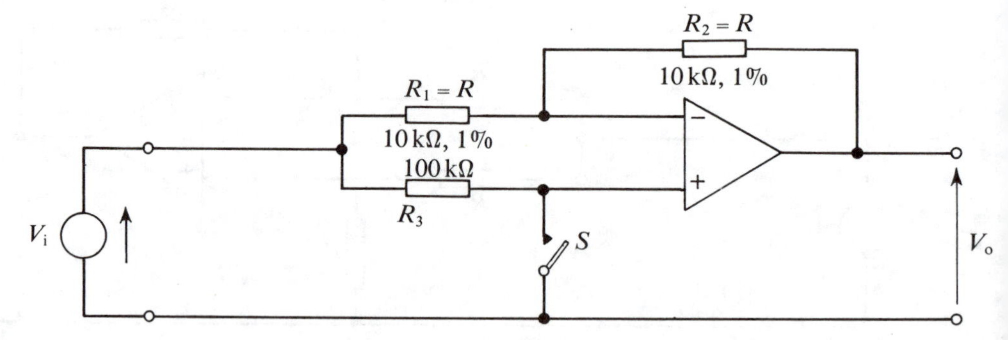

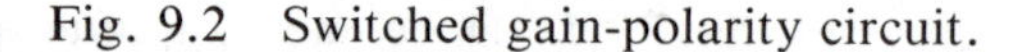
Fig. 9.2 Switched gain-polarity circuit.

With the switch closed R_4 is equal to zero. Substituting this into the voltage transfer function (9.1) gives an indeterminate result because the numerator and denominator expressions both become infinity. This is easily resolved by multiplying the numerator and denominators in (9.1) by the offending R_4 before substituting $R_4 = 0$. This gives

$$A_v = \frac{V_o}{V_1} = \frac{R_4 - \dfrac{R_2}{R_1} \cdot R_3}{R_4 + R_3} = \frac{0 - \dfrac{R}{R} R_3}{0 + R_3} = \frac{-R_3}{+R_3} = -1 \qquad (9.3)$$

Thus the circuit provides a unity gain buffering function in which the output signal is inverted or non-inverted as the switch is closed or open, respectively.

The choice of switch would depend on the application. A mechanical switch, an electrically operated relay contact, or a Field Effect Transistor (FET) switch are among those which could be considered. For high-speed switching an FET switch is to be preferred since it would have a typical operating time of less than 1 μs. This is to be compared with a few milliseconds for a relay switch. Real switches have small but finite on-resistances and very large but not infinite off-resistances. Resistors R_3 and S are connected as a potential divider to the non-inverting op. amp. input terminal. So that the switch action is close to that of an ideal switch, resistance R_3 is chosen to be very large compared with the switch on-resistance and very small compared to the switch off-resistance. In practice, it is usually easy to satisfy both these conditions by ratios of a few orders of magnitude. The main requirement for R_1 and R_2 is that they have closely matched values, $R_1 = R_2 = R$. The accuracy of the inverting gain $A_v = -1$ depends directly on the matching. Typical resistor values are indicated on the figure. When signal amplitudes are low the relative effects of unwanted output voltages due to op. amp. offset voltages and bias currents can be significant. Op. amp. offset V_{os} can be nulled as described in Chapter 7. Bias currents, however, cannot be compensated for, because the resistance presented to the non-inverting op. amp. input is not constant when S is opened and closed. The choice of an op. amp. with very low bias currents (for example an FET input type of op. amp.) usually obviates this problem.

For an explanation of FET switches see Ritchie, G.J. *Transistor Circuit Techniques*, Second edition, Van Nostrand Reinhold (International), 1987.

Offset effects and compensation for them are explained in Chapter 7.

The variable bipolar-gain circuit

In this variant of Fig. 9.1(a) the resistors R_1 and R_2 are set equal and R_3 and R_4 are replaced by a potentiometer R_p as shown in Fig. 9.3. Ignore resistor R_5 for the present. The resistance from the pointer to the bottom of the potentiometer will be some fraction x of R_p and corresponds to R_4. The fraction x varies from 0 to 1 as the pointer is moved from the bottom connection to the top connection of the potentiometer. The upper part of the potentiometer, $R_p - xR_p = (1-x)R_p$ corresponds to R_3. Substituting $R_1 = R_2 = R$, $R_3 = (1-x)R_p$ and $R_4 = xR_p$ into the voltage transfer function (9.1) gives

$$A_v = \frac{V_o}{V_i} = \frac{1 - \dfrac{R}{R} \cdot \dfrac{(1-x)R_p}{xR_p}}{1 + \dfrac{(1-x)R_p}{xR_p}} = \frac{1 - \dfrac{(1-x)}{x}}{1 + \dfrac{(1-x)}{x}} = \frac{x - (1-x)}{x + (1-x)}$$

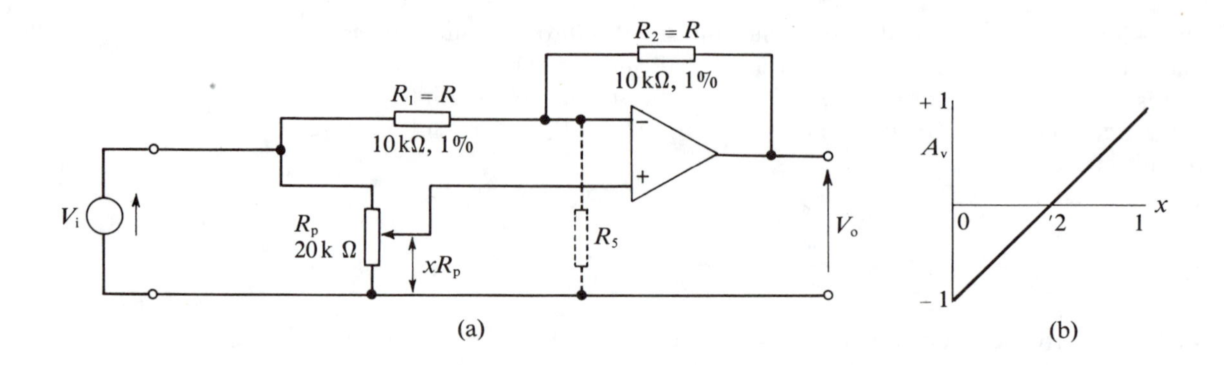

Fig. 9.3 Variable bipolar-gain circuit. (a) Circuit diagram, (b) gain variation with potentiometer setting.

Hence

$$A_v = 2x - 1 \tag{9.4}$$

This function is plotted in Fig. 9.2(b). As the potentiometer is adjusted from low to high the voltage gain varies linearly from $A_v = -1$ to $A_v = +1$. By fitting a pointer knob and a calibrated scale any gain in this range could be set up.

As with the switched gain-polarity circuit, output offset voltages due to bias currents cannot be exactly compensated for because R_p presents a variable resistance to the op. amp. The generally small output offset voltages are more likely to be important relative to the desired component of the output signal when small positive or negative values of A_v are selected. That is when x is approximately 1/2. In this case, the non-inverting op. amp. terminal sees the approximately equal upper and lower halves of R_p connected in parallel, that is $R_p/4$. The inverting op. amp. terminal sees $R_1 \| R_2$, that is $R \| R$ or $R/2$. Hence quite good bias current compensation can be obtained provided $R_p/4 = R/2$ is satisfied. Typical suitable values for the resistors are shown on Fig. 9.3(a).

For some applications a variable gain of wider range than ± 1 is desired, say $\pm N$. One way this can be achieved is to connect the output of the present circuit to a basic non-inverting amplifier (see Fig. 6.9) having $A_v = N$. Alternatively, a gain range of $\pm N$ can be obtained from the present circuit by adding the resistor R_5. Providing the conditions $R_1 = R/N$, $R_2 = R$, $R_5 = R(N-1)$ are satisfied the voltage transfer function is given by

An explanation of how this works can be found in Graeme, J.G. *Designing with Operation Amplifiers: Application Alternatives* (McGraw-Hill, 1977), pp. 22–23.

$$A_v = N(2x - 1) \tag{9.5}$$

The bridge amplifier

A very common application of the bridge amplifier is for use with resistance strain gauge transducers. This type of transducer is often manufactured by etching a pattern in a very thin metal film. The dimensions of the gauge are of the order of 1 cm × 1 cm, and resistance values typically lie in the range 100–1000 Ω. Mechanical stretching of the gauge causes a proportionate increase in its resistance from R to $R + \delta.R$. In practice the fractional change in resistance δ is very small. The attractiveness of the strain gauge arises from the variety of mechanical arrangements which can be used with the gauge to measure quantities such as force, strain, pressure, and acceleration. The combination of transducer and bridge amplifier provides an output voltage which is proportional to the physical quantity being measured. This output voltage can then be used to drive a meter, data logger, etc.

Other types of resistive transducers are controlled by light, temperature, magnetic fields, etc. The bridge amplifier can be used for these cases too.

Consider once again the circuit of Fig. 9.1 and its voltage transfer function as given by Equation 9.1. It is easily verified that $A_v = 0$ if $R_1R_4 = R_2R_3$. This condition is identical to the condition for balance of the well-known Wheatstone bridge. In fact the circuit can be re-drawn to look like the Wheatstone bridge, as shown in Fig. 9.4. A simple and convenient balance condition is to choose all resistors to be of equal value, $R_1 = R_2 = R_3 = R_4 = R$. In this case, the voltage gain $A_v = 0$ and there is no output voltage (assuming the op. amp. to be ideal) even though an input is applied. A small change in any of the four resistances causes the circuit to be unbalanced and an output voltage to appear. Consider placing the transducer of resistance $R(1 + \delta)$ in the R_2-branch, as shown. The voltage transfer function of Equation 9.1 gives

$$A_v = \frac{V_o}{V_i} = \frac{1 - \dfrac{R(1+\delta)}{R} \cdot \dfrac{R}{R}}{1 + \dfrac{R}{R}} = \frac{1 - (1+\delta)}{1+1}$$

Hence

$$V_o = A.V_i = -\frac{\delta}{2} \cdot V_i \tag{9.6}$$

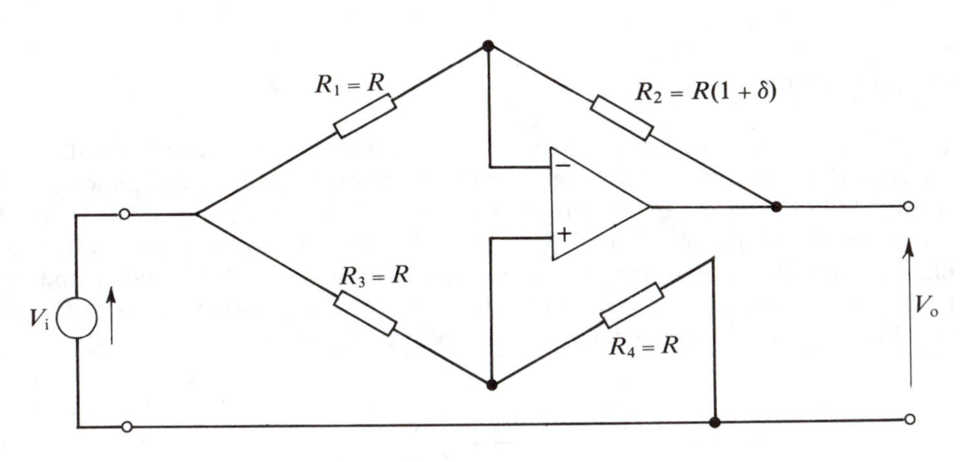

Fig. 9.4 Bridge amplifier.

This demonstrates the linear relationship between change of resistance and output voltage. Voltage V_i is required to have a known and stable value.

The output voltage when the gauge is inserted in each of the four branches is given by the following relationships:

As an exercise, derive these relationships yourself.

Gauge in R_1 branch

$$V_o = +\frac{\delta}{2} \cdot \frac{1}{1+\delta} \quad V_i \approx +\frac{\delta}{2} \cdot V_i$$

Gauge in R_2 branch [from (9.6)]

$$V_o = -\frac{\delta}{2} \cdot V_i$$

Gauge in R_3 branch (9.7)

$$V_o = -\frac{\delta}{2} \cdot \frac{1}{1+\delta/2} \quad V_i \approx -\frac{\delta}{2} \cdot V_i$$

Gauge in R_4 branch

$$V_o = +\frac{\delta}{2} \cdot \frac{1}{1+\delta/2} \quad V_i \approx +\frac{\delta}{2} \cdot V_i$$

Only when the gauge is in the R_2 branch is the relationship exactly linear. However, because in practice $\delta \ll 1$ the linearity is good with the other gauge locations. Where two gauges can be mechanically mounted together they can be connected in opposing branches of the bridge (say in the R_2 and R_3 branches) to double the sensitivity.

Prove this yourself.

Even with a doubling of the sensitivity a further amplifier stage may be needed. Any unwanted offset voltages at the output of the bridge amplifier would also be amplified and could lead to unacceptable errors. This could be avoided by using an alternating source for V_1 rather than a constant one as shown. The output of the bridge circuit is then connected to an a.c. amplifier, which amplifies the desired alternating component present in V_o but not the unwanted constant component due to output offset voltages. A.C. amplifiers are discussed later in this chapter.

The phase shifter

The circuits derived so far from the precision difference amplifier have all used resistors with the op. amp. However, circuits can be explored using any impedances. In the phase shifter a capacitor is used in place of one of the resistors. A phase shifter has the properties of constant gain magnitude but variable phase-shift. One phase shifter circuit is shown in Fig. 9.5(a). Its behaviour can be understood by reference once again to the voltage transfer function of Equation 9.1, where R_4 will now be replaced by the impedance of the capacitor C_4. Thus

$$\hat{A}_v = \frac{\hat{V}_o}{\hat{V}_i} = \frac{1 - \frac{R}{R} \cdot \frac{R_3}{1/(j\omega C_4)}}{1 - \frac{R_3}{1/(j\omega C_4)}} = \frac{1 - j\omega C_4 R_3}{1 + j\omega C_4 R_3} \tag{9.8}$$

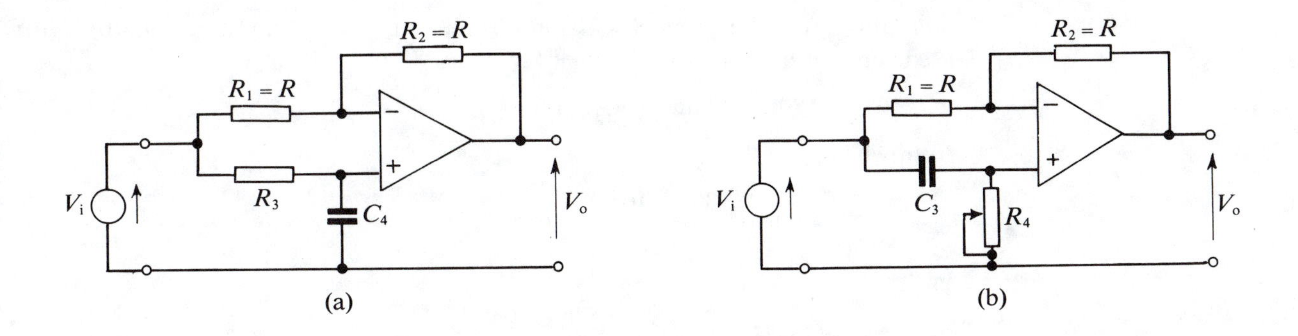

Fig. 9.5 Phase shifter: (a) capacitor in branch four, (b) capacitor in branch three.

In phasor form $\hat{A}_v = |\hat{A}_v|\underline{/\phi}$, where from Equation 9.8

$$|\hat{A}_v| = \frac{\sqrt{[1 + (-\omega C_4 R_3)^2]}}{\sqrt{[1 + (\omega C_4 R_3)^2]}} = \frac{\sqrt{(1 + \omega^2 C_4^2 R_3^2)}}{\sqrt{(1 + \omega^2 C_4^2 R_3^2)}} = 1$$

Thus $\hat{A}_v$ has constant magnitude as required. The phase angle formula is obtained as follows:

$$\phi = \text{(angle of numerator of Eqn 9.8)} - \text{(angle of denominator of Eqn 9.8)}$$
$$= \tan^{-1}\left(\frac{-\omega C_4 R_3}{1}\right) - \tan^{-1}\left(\frac{\omega C_4 R_3}{1}\right)$$
$$= -\tan^{-1}\left(\frac{\omega C_4 R_3}{1}\right) - \tan^{-1}\left(\frac{\omega C_4 R_3}{1}\right)$$

Thus

$$\phi = -2\tan^{-1}(\omega C_4 R_3) \tag{9.9}$$

Phase angle ϕ is determined by the choice of C_4 and R_3. For variable phase angle, a variable resistor R_3 can be used as shown in Fig. 9.5(a). With R_3 set to the minimum of $R_3 = 0$, then $\phi = -2\tan^{-1}(0) = 0$. As R_3 is increased the phase angle increases negatively towards the limiting value of $-180°$. This limit can only be approached since to reach it would require R_3 to increase to infinity. Practical variable resistors always have some finite upper limit.

For all except very small valued variable capacitors it is usually found that variable resistors are cheaper.

An alternative phase shifter is shown in Fig. 9.5(b). In this circuit the third branch is a capacitor. The gain modulus and phase are given by $|\hat{A}| = 1$, and

$$\phi = 180° - 2\tan^{-1}(\omega C_3 R_4) \tag{9.10}$$

Derive this yourself.

In this circuit using a variable R_4 provides a positive phase angle which varies from a lower value of 0° in the limit to an upper value of +180°. The lower limit is approached but not reached as R_4 reaches a finite maximum.

Worked Example 9.1

Design a phase shifter to operate on 1 kHz sinusoidal signals having a maximum voltage r.m.s. amplitude of 5 V. The phase is to be variable from +10° to +180°.

Solution: From the above discussion it is seen that the circuit in Fig. 9.5(b) is to be used. From Equation 9.10 for $\phi = 10°$,

$$10° = 180° - 2\tan^{-1}(\omega C_3 R_4)$$

from which

$$\omega C_3 R_4 = \tan\left(\frac{180° - 10°}{2}\right) = 11.43$$

Thus

$$C_3 R_4 = \frac{11.43}{\omega} = \frac{11.43}{2\pi \times 1 \text{ kHz}} = 182 \times 10^{-3} \text{ s}$$

Capacitors and variable resistors are manufactured to have preferred values. Preferred-value components will be chosen here in preference to special-valued components because of the lower cost. The requirement is to choose C_3 and R_4 values so that their product equals or exceeds 1.82×10^{-3} seconds to ensure meeting the 10° lower limit on phase shift. Extremes of resistor values should be avoided. Very low values will draw large current and very high values can lead to high-frequency effects due to interaction with the inevitable stray capacitances, as well as unwanted increases in output offset voltages. Acceptable resistances typically lie in the range 10^2–10^5 Ω. In this application the actual value chosen is not critical. A study of suppliers' catalogues shows that the requirement is met by $C_3 \times R_4 = 0.22\ \mu\text{F} \times 10\ \text{k}\Omega = 2.2 \times 10^{-3}$ seconds. Because R_4 is variable, bias current compensation cannot be exactly obtained using fixed R_1 and R_2 resistors. A compromise is to compensate for the mid-value of R_4, that is 5 kΩ. This requires $R_1 = R_2 = 10$ kΩ to be chosen. Good matching of R_1 and R_2 (to within 1% say) is needed to obtain accurate unity gain-modulus.

Other values for C_3 and R_4 can be readily found which meet the requirements. You might like to do this.

A general purpose low-cost op. amp. such as the 741 will meet the specification. Using Equation 7.19 of Worked Example 7.2 in Chapter 7, a maximum output voltage slope of $2\pi f V_{om} = 2\pi \times 10^3 \times 5 \times \sqrt{2} = 4.4 \times 10^4$ V/s is expected. Manufacturers' data for the 741 op. amp. gives a minimum slew rate of 0.5×10^6 V/s, so slew rate distortion will not be a problem. Power supply voltages need to be chosen for the 741. The gain modulus $|\hat{A}|$ is unity and so the maximum op. amp. output will be 5 V r.m.s. Thus the peak output swing will be $\pm 5 \times \sqrt{2} = \pm 7.07$ V. The op. amp. input terminals could have a voltage swing up to ± 7.07 V. For the op. amp. to work properly the power supplies need to be chosen to encompass these swings by a couple of volts. Commonly available supply voltages of ± 12 V are a suitable choice and Figs 6.1(b) and 6.11 show how to connect them to the op. amp.

In summary, the components are

resistors,	$R_1 = R_2 = 10$ kΩ(1%)	op. amp:	741 type
	$R_4 = 10$ kΩ variable	power supplies:	± 12 V
capacitor,	$C_3 = 0.22\ \mu$F		

A.C. Amplifiers

Where the signal to be amplified is alternating with some lower bound on frequency, there can be advantages in using an a.c. amplifier. An a.c. amplifier only amplifies

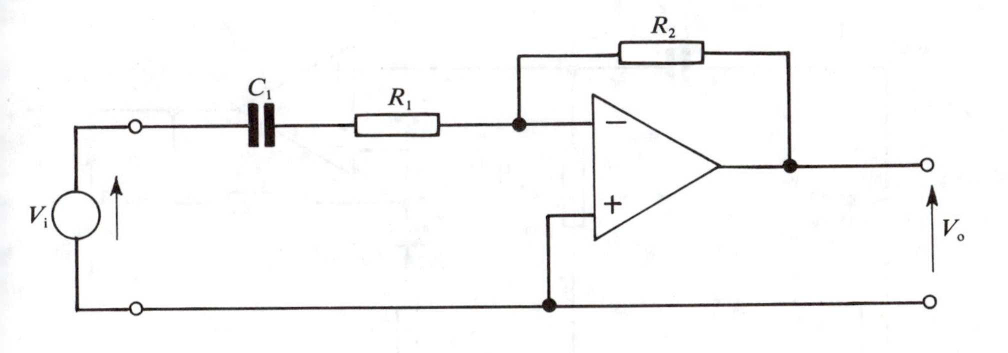

Fig. 9.6 Inverting a.c. amplifier.

alternating components (a.c.) and does not respond to direct components (d.c.). This property can be useful, for example where the signal contains an unwanted direct component as from some transducers or perhaps as a result of output offset voltages from previous stages of op. amp. circuits.

A simple inverting a.c. amplifier is shown in Fig. 9.6. This can be recognized as the basic two-resistor inverting voltage amplifier of Fig. 6.4 with a capacitor C_1 augmenting R_1 in series. The voltage transfer function is therefore given by $A_v = -R_2/R_1$ with R_1 replaced by the impedance of C_1 in series with R_1. Hence

$$A_v = -\frac{R_2}{\dfrac{1}{j\omega C_1} + R_1} = -\frac{R_2}{R_1}\,\frac{1}{1 + \dfrac{1}{j\omega C_1 R_1}} \tag{9.11}$$

This relationship takes the form of a mid-band gain term $-R_2/R_1$ multiplied by a factor due to a low-frequency effect. The lower half-power frequency is given by

See Chapter 3 for a discussion of low-frequency effects.

$$f_L = \frac{1}{2\pi C_1 R_1} \tag{9.12}$$

This frequency denotes the region below which the voltage gain falls away. At zero frequency (that is the frequency of any d.c. component of signal) $\omega = 0$ and Expression 9.11 indicates a gain of zero. Physically what happens in the circuit is that once the capacitor has initially charged up to any d.c. voltage present, no further direct current flows. The capacitor appears as an open circuit to the direct components and the amplifier does not respond to them. A capacitor used in this way is called a *d.c. blocking* capacitor.

Direct components may still appear at the op. amp. output due to op. amp. voltage offsets and bias current effects. These can be nulled and compensated as explained in Chapter 7. However, for a.c. signals they can be ignored as long as the output offset is not so large as to significantly affect the available output voltage swing. The unwanted d.c. output offset can be eliminated by a blocking capacitor connected in series with the signal path at the circuit output. A blocking capacitor is not required here if a further a.c. op. amp. circuit follows since that circuit does not respond to a d.c. input.

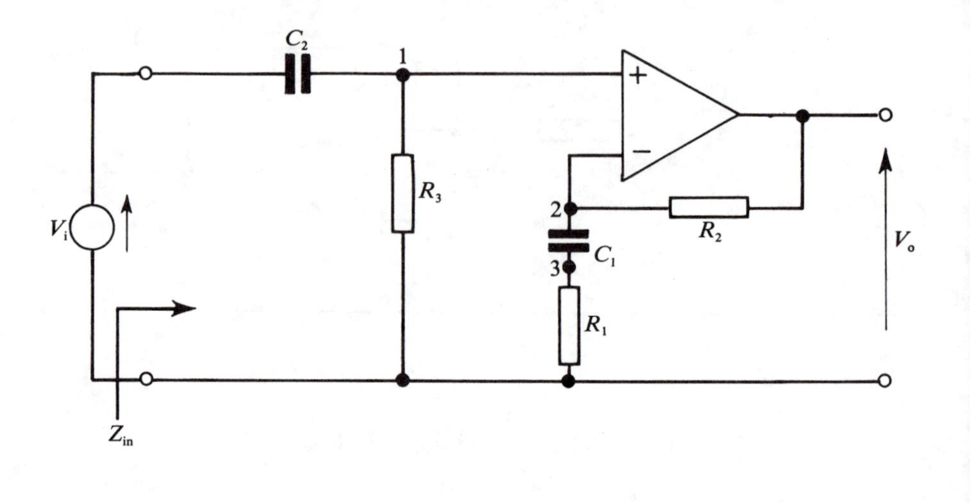

Fig. 9.7 Non-inverting a.c. amplifier.

For the basic non-inverting amplifier refer to Chapter 6 and the discussion accompanying Fig. 6.9.

An a.c. signal version of the basic non-inverting amplifier is shown in Fig. 9.7. Capacitor C_1 is chosen to have negligibly small reactance at signal frequencies and therefore the gain is set by resistors R_1 and R_2 and given by the formula, $A_v = 1 + (R_2/R_1)$. This may be confirmed by consulting Fig. 6.9 and Equation 6.31, and noting that R_1 and R_2 have been interchanged in Fig. 9.7. Blocking capacitor C_2 is inserted in the input signal path to prevent any d.c. input from being amplified. The value of C_2 is chosen to give a reactance at signal frequencies which is very much less than R_3. Resistor R_3 is included to provide a path for the op. amp. bias current I_b^- to flow. Without the resistor the circuit would not function because of the d.c. blocking effect of C_2. As with the previous circuit, output offsets can occur due to V_{os} and the bias currents of the op. amp. Output offsets are worse with high d.c. gains than with low. Without capacitor C_1 inserted in series with R_1 the d.c. gain ratio would be the same as the gain at signal frequencies. For some designs the gain needs to have a high value. Where this can lead to unacceptable output offset voltages, the inclusion of C_1 can help. The presence of C_1 causes the d.c. resistance of the branch containing R_1 to be infinity and so the d.c. voltage gain is reduced to unity: $A_v = 1 + (R_2/\infty) = 1$.

The d.c. voltage gain referred to here is for the part of the circuit to the right of blocking capacitor C_2.

Capacitors C_1 and C_2 cause two low-frequency effects located respectively at

$$f_{L1} = \frac{1}{2\pi C_2 R_3} \text{ and } f_{L2} = \frac{1}{2\pi C_1 R_1} \tag{9.13}$$

For an ideal op. amp. the input impedance is infinity.

One of the reasons for using the non-inverting amplifier rather than the inverting amplifier is that the former has very high input impedance. This feature is required to avoid loading effects when the signal source has high self impedance (as for example with a piezo-electric microphone). The need to include R_3 in the non-inverting a.c. amplifier, Fig. 9.7, causes a lower input impedance, $Z_{in} = R_3$ to be presented to the input source at signal frequencies. This unwanted effect is avoided in the circuit modification shown in Fig. 9.8, where the lower terminal of resistor R_3

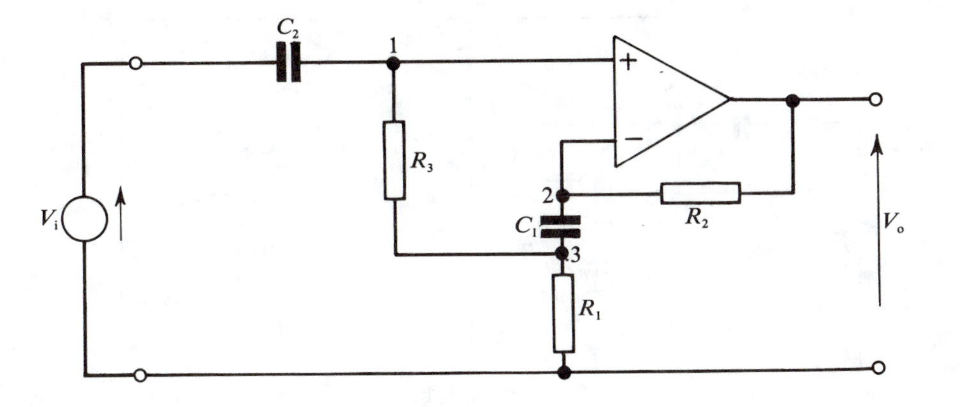

Fig. 9.8 Bootstrapped non-inverting a.c. amplifier.

is now connected to node 3. To understand how this works refer back to the unmodified circuit, Fig. 9.7. Due to the normal negative feedback action of the very high gain op. amp., the signal voltages at nodes 1 and 2 are identical or very nearly so. Capacitor C_1 is chosen to have low reactance at signal frequencies, thus the a.c. signals at nodes 2 and 3 are practically identical. Hence any a.c. signal at node 1 will appear at node 3 with the same amplitude and phase. By connecting resistor R_3 across these two nodes as in Fig. 9.8, no a.c. voltage will appear across R_3 and thus no a.c. signal current will be drawn from the input signal source, thus providing an infinite input impedance (or nearly so for non-ideal op. amp. and finite C_1). This action is a beneficial form of positive feedback. The a.c. voltage at the top of R_3 is fed back in phase at node 3 and the lower terminal of R_3 is 'pulled up by its own bootstraps'. This technique is called *bootstrapping*.

Most op. amp. circuits are constructed using dual power supplies; one for V^+ and one for V^-. There are occasions where *single-power-supply* operation is an advantage. For example, this can save on batteries in portable apparatus or can be justified by the cost savings obtained from reducing the number of electronic power supply circuit modules from two to one. Single-power-supply operation is possible with a.c. amplifiers because of the d.c. isolation property of the blocking capacitors inserted in the signal paths. On the amplifier side of the blocking capacitors it does not matter to the signal what the steady-state d.c. signal voltages are, since these are removed by the blocking capacitors. The only requirement is that whatever voltages are chosen for the op. amp. power supplies, the op. amp. input and output terminals must always have voltages which lie within the range V^-–V^+. Hence successful single-supply operation can be obtained with $V^- = 0$ provided the op. amp. terminal voltages are raised to lie within the range 0–V^+. To obtain the same output signal swing as for dual supplies the single-supply voltage should now equal the sum of the dual-supply voltage magnitudes.

For example, a dual supply of ± 12 V is replaced by a single supply of 24 V.

Fig. 9.9(a) shows how this idea can be implemented for a non-inverting a.c. amplifier. The op. amp. V^- and V^+ are connected to the zero volts line and the single-supply V_{SS}. Under no-signal conditions equal resistors R_3 and R_4 provide a

Resistors R_3 and R_4 also provide the necessary d.c. path for bias current I_b^+.

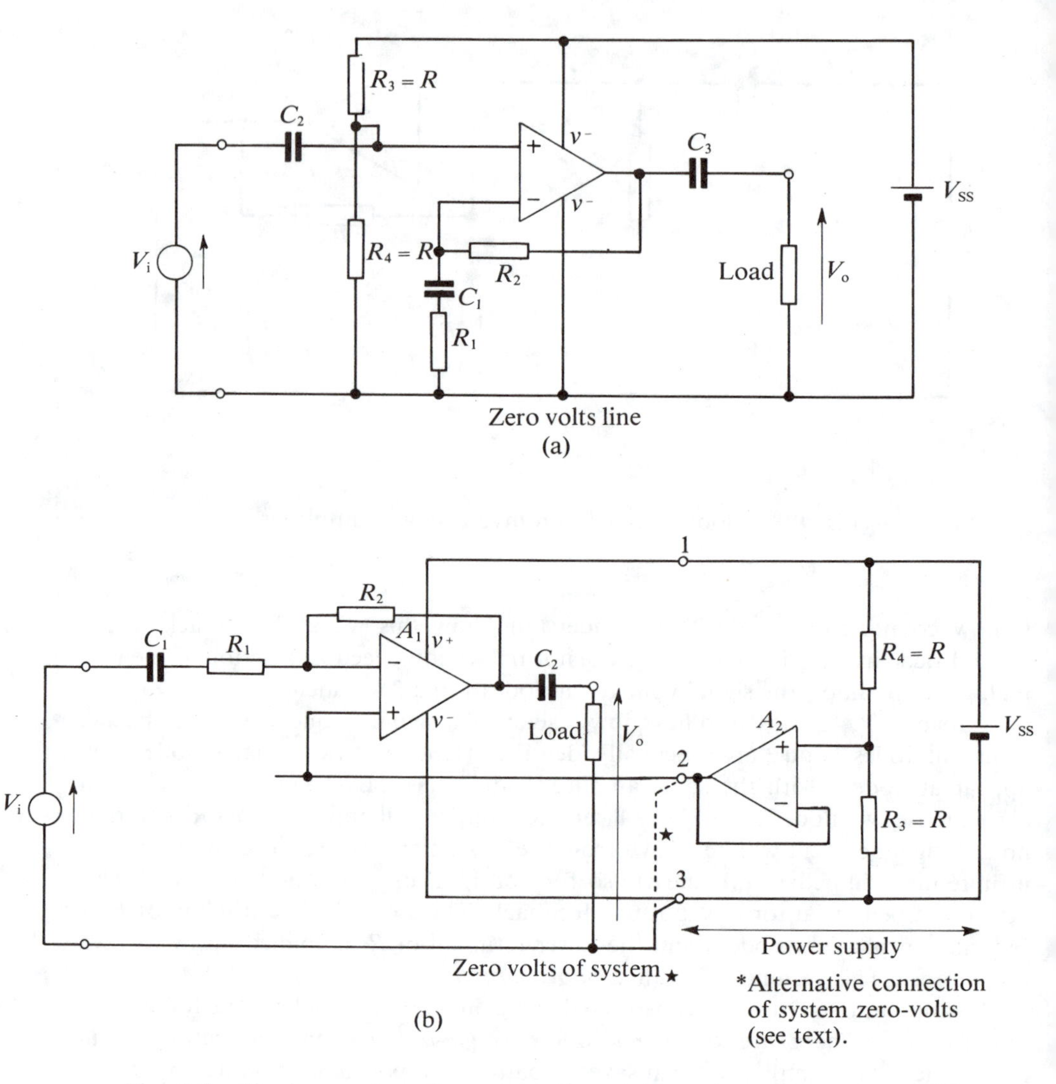

Fig. 9.9 Single-supply voltage operation: (a) non-inverting a.c. amplifier, (b) inverting a.c. amplifier.

d.c. voltage of $\frac{1}{2} V_{SS}$ to the non-inverting op. amp. input. The presence of C_1 causes the op. amp. feedback circuit to have a d.c. gain of unity as previously explained. Hence all terminals of the op. amp. will lie within the voltage supply range $0-V_{SS}$ volts as required. It will usually be essential to have capacitor C_1. Without it the circuit has a d.c. voltage gain of $1 + R_2/R_1$ and the op. amp. output could be driven into saturation by the $\frac{1}{2} V_{SS}$ volts at the non-inverting op. amp. input terminal. Capacitor C_3 serves to block the d.c. op. amp. output from the load. When signals are applied, normal a.c. amplification takes place and the voltages at the op. amp. terminals fluctuate about the average value of $\frac{1}{2} V_{SS}$.

Single-supply operation for inverting a.c. amplifiers can be obtained using the arrangement in Fig. 9.9(b). The requirement that the d.c. voltages at the op. amp.

terminals be moved into the range 0–V_{SS} volts is again achieved by driving the op. amp. non-inverting terminal at ½ V_{SS} from a potential divider, R_3 and R_4 across the single supply. Strictly the unity buffer circuit A_2 is not required since only a tiny bias current I_b^+ is drawn from it by A_1. However, it can play a useful role in the following circumstances. Where V_{SS} is fully 'floating' (as in a battery supply for example) the location of the system zero-volts point may be chosen freely. It can therefore be moved from terminal 3 to terminal 2. The effect is to drive V^+ to + ½ V_{SS} and V^- to – ½ V_{SS}. Conventional dual supplies are therefore restored. Hence *any* op. amp. circuits (not just the a.c. amplifier indicated) may now be used with V^+ and V^- connected to terminals 1 and 3 respectively and the circuit zero-volts points connected to terminal 2. In general, the current through terminal 2 will not now be tiny and unity buffer A_2 prevents current loading on the R_3 and R_4 potential divider.

Worked Example 9.2

Design a non-inverting a.c. amplifier to have a mid-band voltage gain of 100 and an input impedance of 1 MΩ. Both low-frequency cut-offs are to be at 10 Hz maximum.

Solution: The circuit in Fig. 9.7 is to be designed. The mid-band input impedance is given by $Z_{in} = R_3$ and it follows that $R_3 = 1$ MΩ is chosen, which is a preferred value. Capacitor C_2 must satisfy $f_{L2} = 1/(2\pi C_2 R_3)$; that is $C_2 = (2\pi f_{L2} R_3)^{-1} = (2\pi \times 10 \text{ Hz} \times 1 \text{ M}\Omega)^{-1} = 15.9$ nF. To ensure the cut-off frequency is not greater than 10 Hz a preferred value above this is chosen. A convenient preferred value is $C_2 = 22$ nF.

Resistors R_1 and R_2 are chosen to satisfy $A_v = (1 + R_2/R_1) = 100$. By choosing $R_2 = R_3 = 1$ MΩ bias current compensation can also be obtained because the op. amp. inverting and non-inverting terminals see d.c. resistances of R_2 and R_3 respectively. For $A_v = 100$, the choice of $R_2 = 1$ MΩ, requires that $R_1 = 9.9$ kΩ. Assuming the application can tolerate the slight deviation in A_v obtained, the nearby preferred value of $R_1 = 10$ kΩ is chosen. This leads to $A_v = 101$.

Lastly, C_1 is obtained to satisfy $f_{L1} = (2\pi C_1 R_3)$, that is $C_1 = (2\pi f_{L1} R_3)^{-1} = (2\pi \times 10 \text{ Hz} \times 10 \text{ k}\Omega)^{-1} = 1.59\ \mu$F. A nearby preferred value is 2.2 μF.

Design summary: $R_1 = 10$ kΩ, $R_2 = R_3 = 1$ MΩ, $C_1 = 2.2\ \mu$F, $C_2 = 22$ nF.

Active Filters

The subject of active filters touched on in the previous chapter is now taken up in greater detail. However, before doing so some remarks about filters in general are made. Signals can comprise many components having different frequencies. A circuit whose transfer function treats the amplitudes and phases of some frequency components differently from those of others is a *filter*. This very general definition allows the inclusion of circuits having amplitude and phase responses versus frequency of any shape. However, certain types of responses are more commonly met than others and include the *low-pass (LP)*, *high-pass (HP)*, and *band-pass (BP)* amplitude responses. The ideal forms of these responses are graphed in Fig. 9.10. Each has a *pass-band* situated at the lower-, upper-, and mid-ranges respectively of

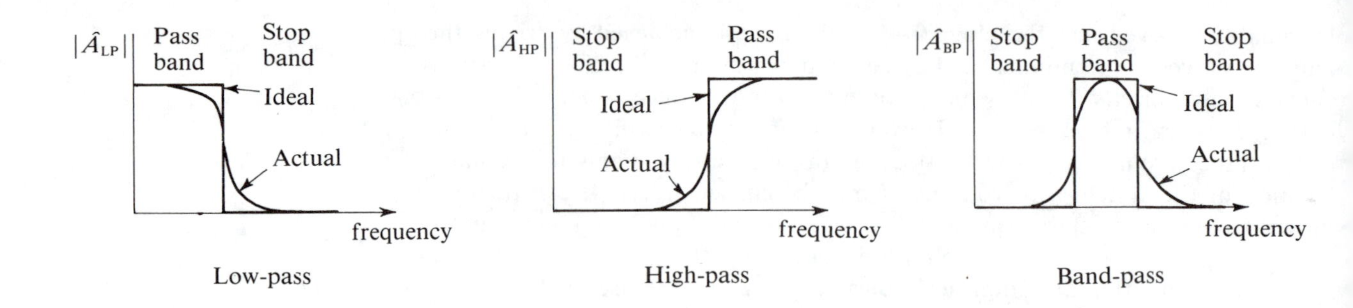

Fig. 9.10 Filter response types.

the frequency axis. Inside the pass-band all frequency signal components are multiplied by a constant transfer function magnitude. Outside the pass-bands the regions are *stop-bands* in which $|\hat{A}| = 0$ and no output at these frequencies occurs. Ideally the pass-band edges are vertical.

Filters are used for many kinds of signal processing applications. An example is to recover the average value (that is the d.c. or zero-frequency component) from a fluctuating signal. A low-pass filter could be used here in which the cut-off frequency is chosen to be low enough to place all the fluctuating component frequencies in the stop-band. Conversely, in another application if the fluctuating components are desired but the d.c. content is not (perhaps because it is an unwanted offset voltage) then a high-pass filter would be used. A band-pass filter could be employed to condition a speech signal for transmission through a 300–3400 Hz telephone channel.

It follows that the a.c. amplifier is a kind of high-pass filter.

In practice these ideal responses can only be approximated. Fig. 9.10 also shows responses that might be obtained from typical real filter circuits. Qualities of a good practical filter response include flatness of pass-band, sharpness of pass-band edges, steepness of the transition from pass-band to stop-band, and high attenuation in the stop-band. As a general rule, filters with responses closer to the ideal are more complex and costly.

Historically, the theory of filters was developed for applications in communications. These filters were typically *LC filters* constructed from inductors and capacitors. Such designs are still successful and economically viable for pass-bands covering the higher frequency ranges (typically in the region of 500 kHz and above). However, designs covering lower frequency ranges require increasingly larger inductor and capacitor values. The inductors in particular soon cause difficulties because of their size and cost. LC filters are not really a practical design option for applications in the low audio frequency range.

In more recent times high performance active devices such as op. amps. have become readily available. These devices can be employed with resistors and capacitors to make filters (called *active filters*) which achieve the desired response at low frequencies without the need for inductors. Active filters now find wide application in communications, instrumentation and many other fields.

The *first-order* filter is the simplest that can be obtained. The general form of its transfer function is

$$\hat{A} = \frac{\hat{V}_o}{\hat{V}_i} = \frac{b_0 + b_1.(j\omega)}{1 + a_1.(j\omega)} \tag{9.14}$$

The *order* equals the highest power of ω found in the transfer function.

Setting b_1 and b_0 equal to zero in turn gives transfer functions having low-pass and high-pass responses,

$$\hat{A}_{LP} = \frac{b_0}{1 + a_1.(j\omega)} \tag{9.15}$$

and

$$\hat{A}_{HP} = \frac{b_1.(j\omega)}{1 + a_1.(j\omega)} \tag{9.16}$$

The forms of the magnitude responses $|\hat{A}_{LP}|$ and $|\hat{A}_{HP}|$ are shown in Fig. 9.11. Consider the low-pass response $\hat{A}_{LP}$. For high frequencies at which $1 \ll a_1.(j\omega)$, the denominator of (9.15) is approximately $a_1.(j\omega)$ and $|\hat{A}_{LP}| \approx b_0/(a_1\omega)$. This is inversely proportional to ω, which on the log-scales of Fig. 9.11 appears as a region which falls linearly into the stop band. The slope of this linear asymptote is 20 dB for each decade increase in frequency (this is written as -20 dB/decade). This and the horizontal asymptote meet at the corner frequency, ω_0. A similar discussion is applicable to the high-pass response of Fig. 9.11(b).

If one frequency is ten times another they are separated by a *decade*.

From the viewpoint of a filter application, the main response features of impact are the location of the band-edge corner frequency at ω_0 and the asymptotic value of transfer function modulus in the pass band, H. It is a simple matter to derive the formulae shown in Table 9.1 relating transfer function coefficients to H and ω_0.

Check this yourself.

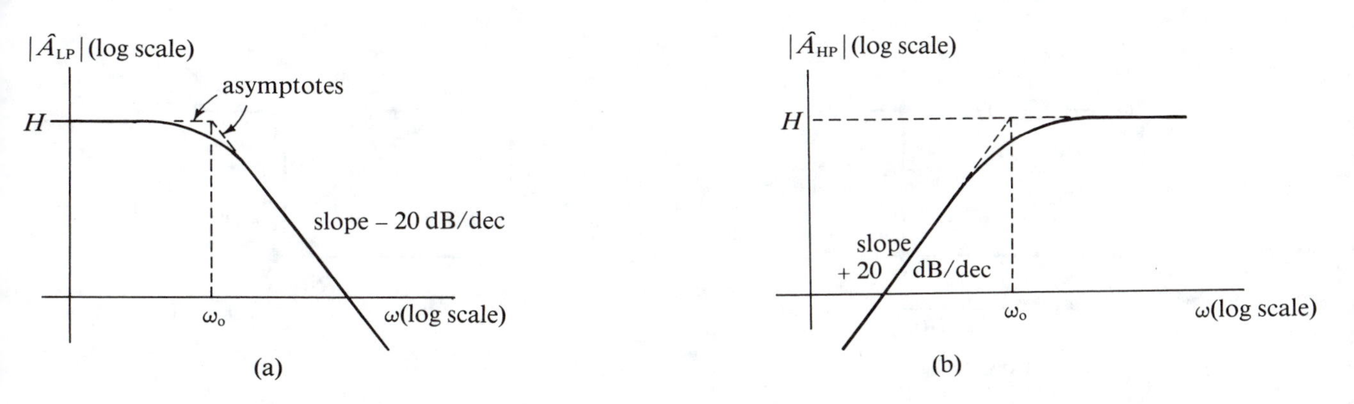

Fig. 9.11 First-order filter responses: (a) low-pass; (b) high-pass.

Table 9.1 First-order Transfer Function Coefficients from Response Features

Low-pass	High-pass
$\hat{A}_{LP} = \dfrac{b_0}{1 + a_1.(j\omega)}$	$\hat{A}_{HP} = \dfrac{b_1 \,.\, (j\omega)}{1 + a_1.(j\omega)}$
$b_0 = H$	$b_1 = H/\omega_0$
$a_1 = 1/\omega_0$	$a_1 = 1/\omega_0$

The next task is to find circuits to realise the first-order responses. Two *RC* circuits to do this are shown in Fig. 9.12. It is a simple circuit analysis task to obtain the transfer functions indicated in the figure and a comparison of them with the general low-pass and high-pass forms of Equations 9.15 and 9.16 readily produce the design procedures shown in Table 9.2. The full design follows the sequence: (i) decide on the desired response features; (ii) obtain the transfer function parameters (Table 9.1); (iii) determine the component values. It should be noted that the circuits of Fig. 9.12 cannot realise all possible specifications. The pass-band band gain is constrained to unity, that is $H = 1$. This constraint can be overcome by following the filter with a basic op. amp. inverting or non-inverting amplifier. It is quite often the case that the value of the pass-band gain is not very important. A signal may go through a chain of filtering and other circuits and the desired signal level can be established for all of them by an amplifier at some convenient point.

Here H is not in decibels.

Filters of *second-order* type have improved characteristics over first-order filters. The low-pass and high-pass responses can have sharper corners at the band edge. The responses also fall away more rapidly in the stop-band at a rate of 40 dB/decade. Thus the attenuation increases by 100:1 rather than 10:1 for every decade increase in frequency. In addition, with the second-order transfer function a band-pass response can be obtained.

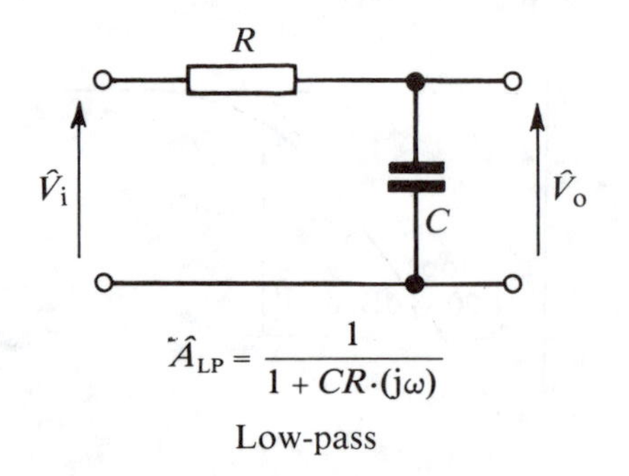

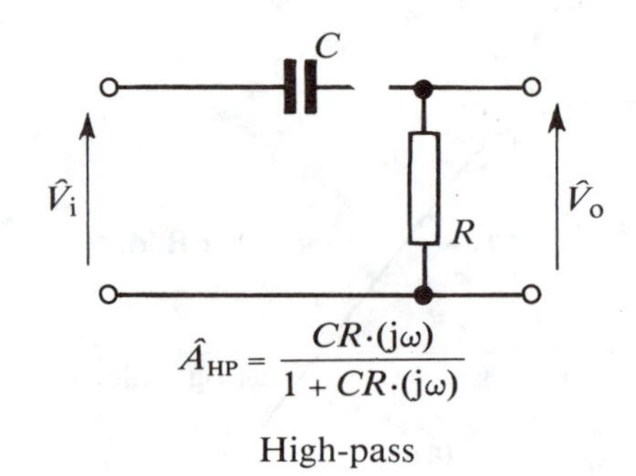

Fig. 9.12 First-order filter circuits.

Table 9.2 Design Procedures for First-Order Filters

Low-pass	High-pass
Choose any C $R = a_1/C$	Choose any C $R = a_1/C$

The general form of the second-order transfer function is:

$$\hat{A} = \frac{V_o}{V_i} = \frac{b_0 + b_1.(j\omega) + b_2.(j\omega)^2}{1 + a_1.(j\omega) + a_2.(j\omega)^2} \tag{9.17}$$

Allowing only one numerator coefficient to exist in turn gives rise to three kinds of response:

Low-pass:

$$\hat{A}_{LP} = \frac{b_0}{1 + a_1.(j\omega) + a_2.(j\omega)^2} \tag{9.18}$$

High-pass:

$$\hat{A}_{HP} = \frac{b_2.(j\omega)^2}{1 + a_1.(j\omega) + a_2.(j\omega)^2} \tag{9.19}$$

Band-pass:

$$\hat{A}_{BP} = \frac{b_1.(j\omega)}{1 + a_1.(j\omega) + a_2.(j\omega)^2} \tag{9.20}$$

These transfer functions have magnitude responses of the forms shown in Fig. 9.13. The 40 dB/decade slope in the stop bands of the low-pass and high-pass responses is shared between both stop bands in the band-pass case which fall away from the peak with slopes approaching 20 dB/decade. A notable feature is the peaks that can occur in the responses with a maximum at or near ω_0. The *Q-factor* of the response is defined as the ratio of the height ($Q.H$) of the curve at ω_0 to the height (H) at the intersection of the asymptotes. For band-pass filters with high Q-values the bandwidth, $\omega_U - \omega_L$, is narrow. It can be shown that $Q = \omega_0/(\omega_U - \omega_L)$. The right-hand term $\omega_0/(\omega_U - \omega_L) = \omega_0/(\text{bandwidth})$ is also called the *selectivity* of the response. This selectivity relationship to Q is also approximately true for the low-pass and high-pass responses. However, it is not usually of interest since these filter responses are not generally used for selecting narrow bands of frequencies by choosing a high Q. Often a relatively flat pass-band response is desired for the low-pass and high-pass responses. The factor Q is then chosen to sharpen up the band edge corner rather than to introduce a peak. An example is the Butterworth response which is maximally flat at $\omega = 0$ and $\omega = \infty$ respectively for the low-pass and high-pass responses. For the second-order case it turns out that maximal flatness is achieved by choosing $Q = 1/\sqrt{2} = 0.7071$.

As with the first-order responses it is not too difficult to derive formulae relating the transfer-function coefficients to the main features of the $|\hat{A}|$ responses (viz, H,

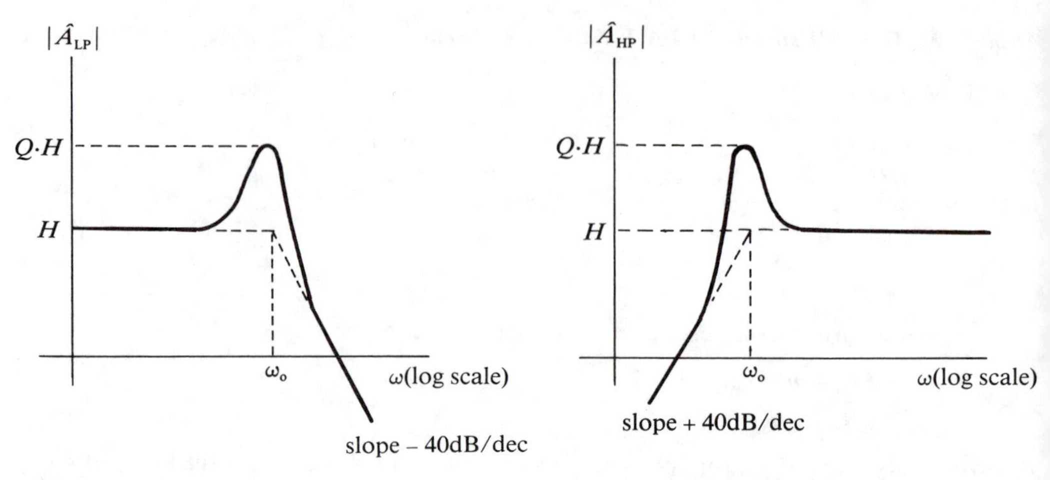

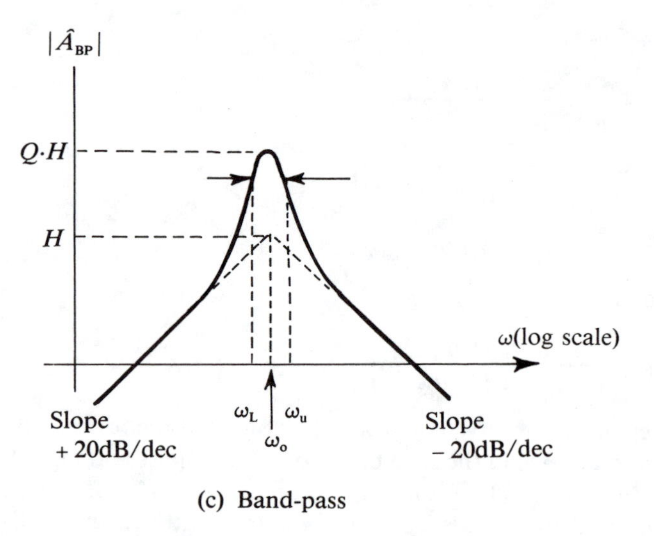

Fig. 9.13 Second-order filter responses: (a) low-pass; (b) high-pass; (c) band-pass.

ω_0 and Q). These formulae are to be found in Table 9.3. Note that the Q-factor only affects coefficient a_1. A low value of a_1 results in a high-Q response and vice versa. Coefficient a_2 determines the corner frequency (ω_0) obtained and H is governed by the value of the appropriate numerator coefficient (b_0, b_1 or b_2). The use of Table 9.3 allows the transfer function coefficients to be calculated from the desired response features.

To implement the transfer function an entirely passive RC circuit should first be considered. A second-order low-pass response can be obtained from cascading two first-order low-pass responses, a second-order high-pass from two first-order high-passes and a band-pass from one first-order low-pass and one first-order high-pass. The second-order responses obtained will have the correct general shapes with the expected attenuation slopes in the stop bands. These low cost solutions are suitable for some applications. Their disadvantage is that only quite low Q-values can be

Table 9.3 Second-Order Transfer Function Coefficients from Response Features

Low-pass	High-pass	Band-pass
$\hat{A}_{LP} = \dfrac{b_0}{1 + a_1.(j\omega) + a_2.(j\omega)^2}$	$\hat{A}_{HP} = \dfrac{b_2.(j\omega)^2}{1 + a_1.(j\omega) + a_2.(j\omega)^2}$	$\hat{A}_{BP} = \dfrac{b_1.(j\omega)}{1 + a_1.(j\omega) + a_2.(j\omega)^2}$
$b_0 = H$	$b_2 = H/\omega_0^2$	$b_1 = H/\omega_0$
$a_1 = \dfrac{1}{Q\omega_0}$	$a_1 = \dfrac{1}{Q\omega_0}$	$a_1 = \dfrac{1}{Q\omega_0}$
$a_2 = 1/\omega_0^2$	$a_2 = 1/\omega_0^2$	$a_2 = 1/\omega_0^2$

obtained (up to about 0.5). For higher Q-values it is required to use an active device such as an op. amp.

A number of second-order filter configurations using one op. amp. have been proposed at various times. A popular one is the *multi-loop feedback* circuit, or Rauch structure. The general form is shown in Fig. 9.14 together with its transfer function. (The derivation of the transfer function is not difficult but is also not very instructive at this point.) By suitable choice of capacitors and resistors for the various branches of this circuit the low-pass, high-pass and band-pass circuits are obtained. These are shown in Fig. 9.15 together with the transfer function formulae obtained by replacing each branch Y_j by $(j\omega C_j)$ or $(1/R_j)$ in the expression in Fig. 9.14 and re-arranging. Each circuit has to provide the desired transfer function as defined by the a and b coefficients in Equations 9.18, 9.19 or 9.20. A direct comparison between them results in conditions which must be met and which are given in Table 9.4.

Note that in Fig. 9.14 admittances (Y) are used which are the reciprocals of impedances ($Z = 1/Y$). This produces a more compact formula in this case.

The design problem is this: given the required transfer function coefficients, find a set of capacitor and resistor values that satisfies the appropriate set of conditions in Table 9.4. There are more unknown components (five) than there are known

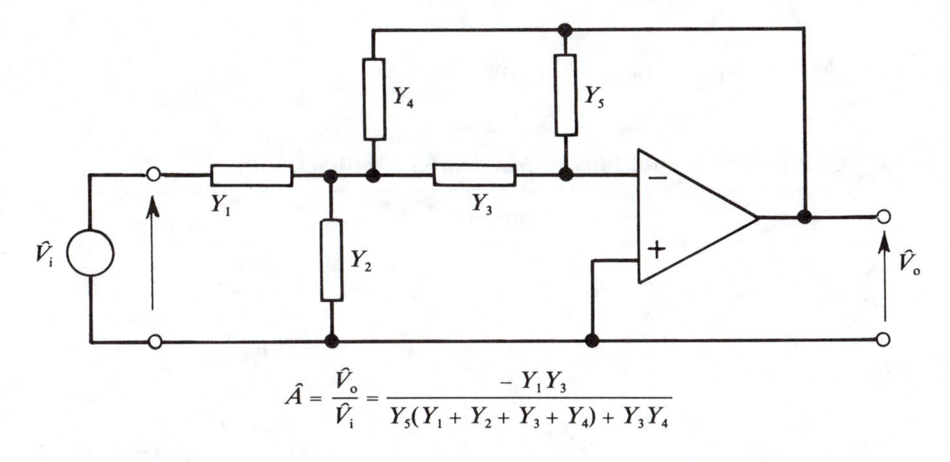

$$\hat{A} = \frac{\hat{V}_o}{\hat{V}_i} = \frac{-Y_1Y_3}{Y_5(Y_1 + Y_2 + Y_3 + Y_4) + Y_3Y_4}$$

Fig. 9.14 General multi-loop feedback active filter circuit.

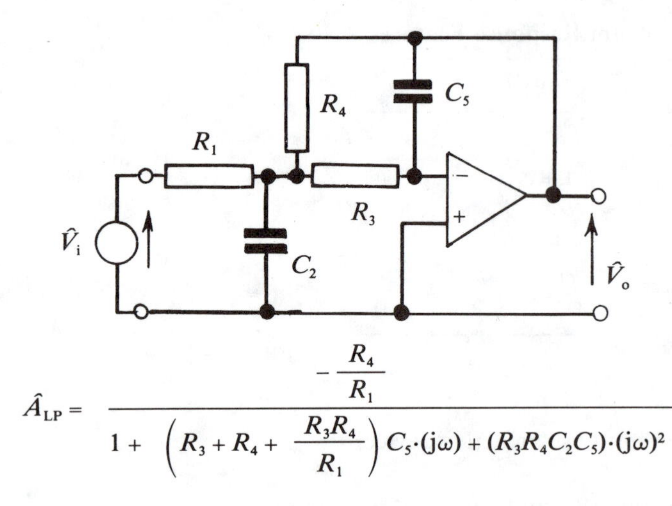

$$\hat{A}_{LP} = \frac{-\dfrac{R_4}{R_1}}{1 + \left(R_3 + R_4 + \dfrac{R_3 R_4}{R_1}\right) C_5 \cdot (j\omega) + (R_3 R_4 C_2 C_5) \cdot (j\omega)^2}$$

Low-pass

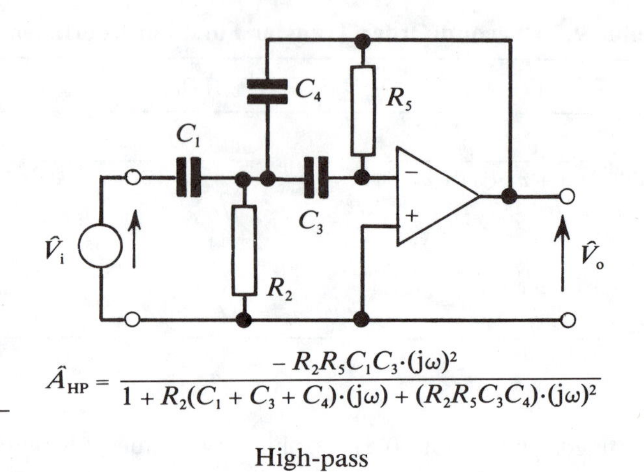

$$\hat{A}_{HP} = \frac{-R_2 R_5 C_1 C_3 \cdot (j\omega)^2}{1 + R_2(C_1 + C_3 + C_4) \cdot (j\omega) + (R_2 R_5 C_3 C_4) \cdot (j\omega)^2}$$

High-pass

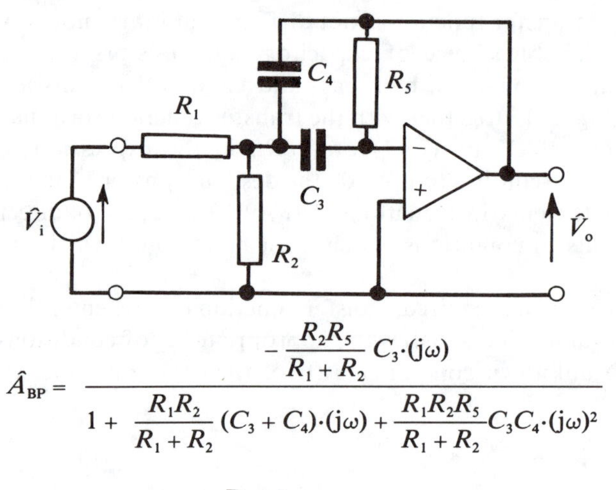

$$\hat{A}_{BP} = \frac{-\dfrac{R_2 R_5}{R_1 + R_2} C_3 \cdot (j\omega)}{1 + \dfrac{R_1 R_2}{R_1 + R_2}(C_3 + C_4) \cdot (j\omega) + \dfrac{R_1 R_2 R_5}{R_1 + R_2} C_3 C_4 \cdot (j\omega)^2}$$

Band-pass

Fig. 9.15 Multi-loop feedback *RC* active filter circuits.

Table 9.4 Component Conditions in Multi-loop Feedback Filters

Low-pass	High-pass	Band-pass
$\dfrac{R_4}{R_1} = -b_0$	$R_2 R_5 C_1 C_3 = -b_2$	$\dfrac{R_2 R_5}{R_1 + R_2} . C_3 = -b_1$
$\left(R_3 + R_4 + \dfrac{R_3 R_4}{R_1}\right) C_5 = a_1$	$R_2(C_1 + C_3 + C_4) = a_1$	$\dfrac{R_1 R_2}{R_1 + R_2}(C_3 + C_4) = a_1$
$R_3 R_4 C_2 C_5 = a_2$	$R_2 R_5 C_3 C_4 = a_2$	$\dfrac{R_1 R_2 R_5}{R_1 + R_2} C_3 C_4 = a_2$

Table 9.5 Design Procedures for Multi-loop Feedback Filters

Low-pass		High-pass		Band-pass	
(i)	Choose convenient R	(i)	Choose convenient C	(i)	Choose convenient C
(ii)	$R_3 = R$	(ii)	$C_1 = C$	(ii)	$C_3 = C$
(iii)	$R_4 = R$	(iii)	$C_3 = C$	(iii)	$C_4 = C$
(iv)	$R_1 = R_4/(-b_0)$	(iv)	$C_4 = C_1 a_2/(-b_2)$	(iv)	$R_1 = a_2/(-b_1 C_4)$
(v)	$C_5 = a_1 // \left(R_3 + R_4 + \frac{R_3 R_4}{R_1}\right)$	(v)	$R_2 = a_1/(C_1 + C_3 + C_4)$	(v)	$R_5 = \frac{C_3 + C_4}{C_3 C_4} . a_2/a_1$
(vi)	$C_2 = a_2/(R_3 R_4 C_5)$	(vi)	$R_5 = -b_2/(R_2 C_1 C_3)$	(vi)	$R_2 = R_1/(C_3 R_5(-b_1) - 1)$

parameters (three) and the solution is not unique. Various design procedures can be formulated depending on what constraints are made to fix the two extra degrees of freedom. One design procedure for each circuit is presented in Table 9.5 where the steps should be followed in the order given. Sometimes impractical component values may result, such as negative or extreme values. In this case new parameters could be tried (adjust H for example), a different design procedure derived, or if that fails, a different circuit tried (such as the biquad which follows the next design example).

The negative signs on the b coefficients result from the inverting property of the circuit. Hence a negative H must be chosen for the circuit to be realisable.

You might be able to find alternative design procedures.

A design follows similar stages to that for first-order filters. First decide on H, ω_0 and Q; then using Table 9.3 determine a_1, a_2 and the b coefficient, and finally calculate the component values using a procedure such as in Table 9.4.

Worked Example 9.3

Design a low-pass second-order multi-loop filter to have a Butterworth response with a corner frequency at 100 rad/s and a voltage gain-magnitude of 2 at zero frequency.

Solution: The Butterworth response requires that $Q = 1/\sqrt{2}$. Also needed are $\omega_0 =$ 100 rad/s and $H = -2$. Note that H is made negative because the multi-loop filter being phase inverting can only provide negative H.

Table 9.3 gives:

$$b_0 = H = -2$$

$$a_1 = 1/(Q\omega_0) = 1.414 \times 10^{-2}\ \text{s}$$

$$a_2 = 1/\omega_0^2 = 10^{-4}\ \text{s}^2$$

The design procedure of Table 9.5 can now be followed. The steps are as follows:

(i) choose a convenient R; say 100 kΩ

(ii) $R_3 = R = 100$ kΩ

(iii) $R_4 = R = 100$ kΩ

(iv) $R_1 = R_4/(-b_0) = 100 \times 10^3/2 = 50$ kΩ

(v) $C_5 = a_1 // \left(R_3 + R_4 + \frac{R_3 R_4}{R_1}\right) = 1.414 \times 10^{-2}/(400\ \text{k}\Omega) = 35.4$ nF

(vi) $C_2 = a_2/(R_3 R_4 C_5) = 10^{-4}/(10^5 \times 10^5 \times 35.4 \times 10^{-9}) = 0.283\ \mu$F

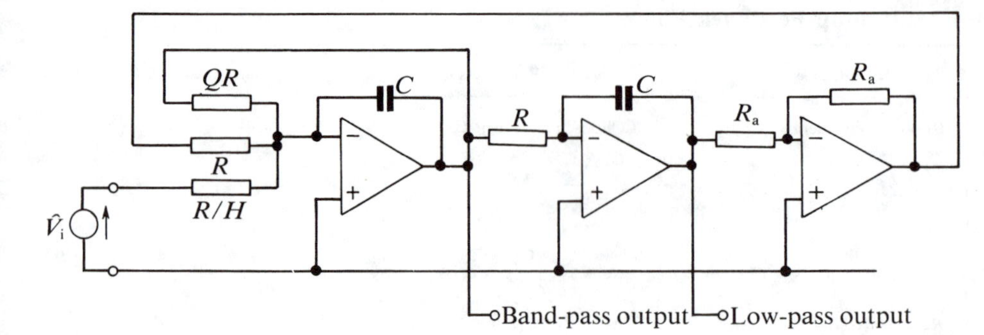

Note: (1) choose R, C to satisfy $RC = \frac{1}{\omega_o}$, (2) choose any convenient value for resistors R_a.

Fig. 9.16 The Tow-Thomas biquad.

The costs of integrated circuit op. amps. are nowadays low enough for filters containing more than one op. amp. to be considered. Second-order circuits of this type include ones derived from analogue computer methods, as is pointed out in the previous chapter. There it is commented that an analogue computer circuit using five op. amps. behaves as an active filter. Using this general approach, circuits with three op. amps. have been found. A popular one is the Tow-Thomas biquad shown in Fig. 9.16. Some advantages of this circuit are worth mentioning. The first is that detailed analysis has shown the circuit has better performance than the single amplifier circuits at high frequencies and at high Q-factors. The second is that the op. amps. provide a degree of isolation between the passive components. Consequently there is less component interaction than with the single amplifier multiloop feedback circuit and the design procedure is thereby much simpler. No recipe of steps is needed; the way to calculate the components can be written directly on the circuit diagram. Lastly the circuit provides simultaneous low-pass, and band-pass outputs.

What of higher-order filters? Higher-order filters are capable of responses which more closely approximate the ideal. Band-edges can be sharper and steeper; stop-bands can have greater attenuation. Stringent response specifications require higher-order filters to be used. One approach is to cascade-up second- and first-order filter sections. Another approach is to transform classical LC filters using active circuits so that physical inductors are avoided. This specialised subject area requires a more detailed study than can be covered here.

Books devoted to filters include Sedra, A.S. and Brackett, P.O. *Filter Theory and Design: Active and Passive* (Pitman, 1979) and Bowron, P. and Stephenson, F.W. *Active Filters for Communication and Instrumentation* (McGraw-Hill, 1979).

Precision Rectifier Circuits

A diode conducts current easily in the forward direction but not in the reverse. Ideally the volts drop in the forward direction, V_F, and the current in the reverse direction, I_R, should both be zero. In real semi-conductor diodes I_R is typically a few micro-amperes and can often be regarded as close enough to the ideal. However, V_F is about 0.6 V and can lead to undesired circuit performance. An example is the basic half-wave rectifier circuit shown in Fig. 9.17(a). Positive polarity inputs are

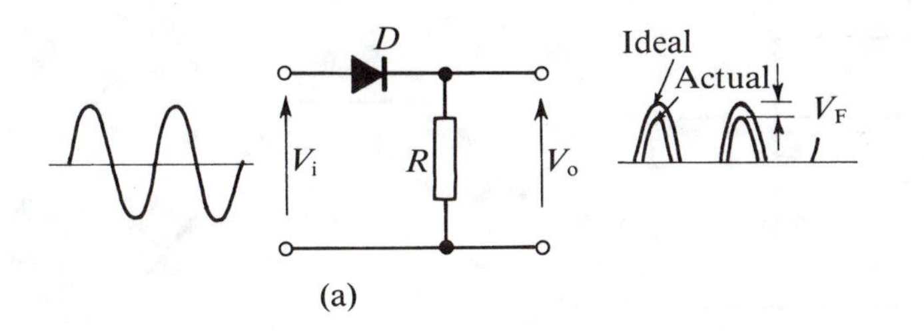

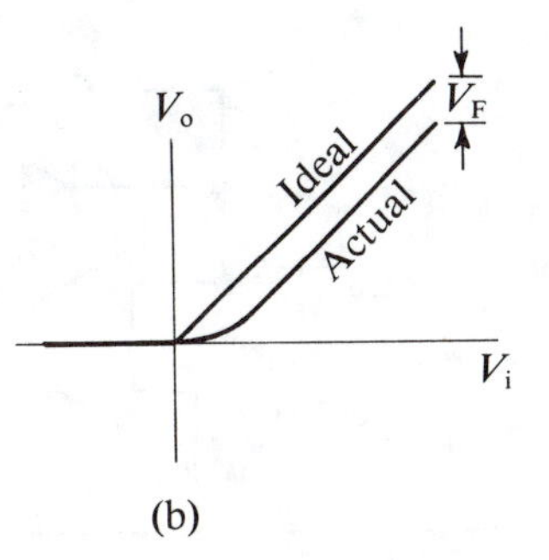

Fig. 9.17 Basic half-wave rectifier circuit: (a) circuit diagram, (b) input–output relationship.

desired to appear unaffected at the output while negative polarities are blocked. With a real diode the rectified signal appears at the output with positive excursions which are less than the ideal by a gap of V_F volts. The input–output characteristics obtained for the circuit are as shown in Fig. 9.17(b). For applications such as a.c./d.c. conversion for a power supply this could be tolerated, but in an instrumentation application an error of this amount usually could not be ignored.

V_F is not constant at 0.6 V but depends on the logarithm of the diode current. Diode characteristics are explained in Sparkes, J.J. *Semiconductor Devices: how they work* (Van Nostrand Reinhold, 1987).

Op. amps. can be used with diodes to overcome these errors by exploiting negative feedback. A precision half-wave rectifier circuit is shown in Fig. 9.18(a) and has the input–output relationship in Fig. 9.18(b). This relationship is seen to conform to the ideal in Fig. 9.17(b) when allowance is made for the output voltage sign reversal which occurs because the op. amp. configuration is inverting.

If desired the output sign could be made to conform exactly with Fig. 9.17(b) by following the half-wave rectifier circuit by a basic inverting amplifier.

Components D_1, D_2 and R_2 appear to the op. amp. as a single composite feedback resistor (albeit a non-linear one) and with input resistor R_1 the circuit is seen to be related to the basic inverting amplifier (Fig. 6.4). Consider the application of a positive input voltage, $V_i > 0$ to the circuit of Fig. 9.18(a). Because of the inverting action of the op. amp. V_a is driven negative. Diode D_1 will be turned on because its right terminal at voltage V_a is negative with respect to its left-hand terminal which is connected via R_2 to the virtual earth at the op. amp. input. This allows current to flow through R_2 as in the basic inverter circuit and so the following precise relationship applies: $V_o = (-R_2/R_1)V_i$. The forward voltage V_F of D_1 is in series with the output impedance of the op. amp. and because of the strong negative feedback in the circuit it has negligible effect on the output voltage. This region of operation corresponds to the downward sloping portion of Fig. 9.18(b).

Note that unusually the op. amp. output at V_a is not the output of this circuit. The output is between R_2 and D_1.

Now consider a negative input voltage, $V_i < 0$. In this case V_a goes positive until at about $V_a = 0.6$ V D_2 turns on to provide a current path for the current flowing through R_1 and into the negative V_i. Since V_a is positive the other diode, D_1, will be off and no current will flow through R_2. The circuit output taken from R_2 is therefore at zero volts. Thus any negative input voltage is blocked from appearing at the output. This corresponds to the horizontal portion of the characteristic in Fig. 9.18(b).

Actually output V_o will not be at exactly zero volts. In addition to any offset voltage contributed by the op. amp. there will be a component $(R_2/_R)$ volts due to the small reverse current of the diode. Usually this can be made acceptably small.

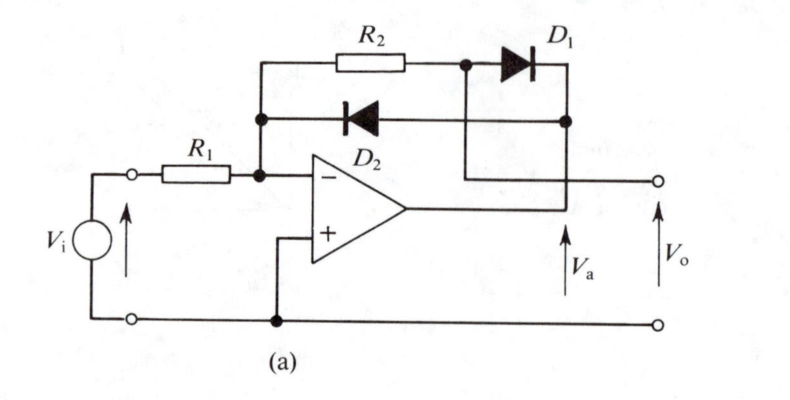

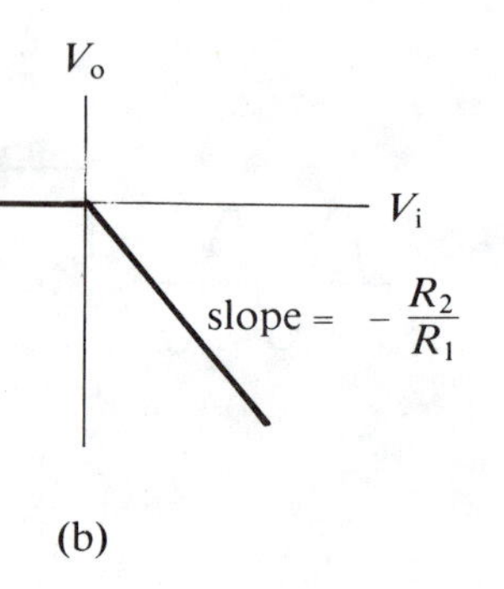

Fig. 9.18 Precision half-wave rectifier: (a) circuit diagram, (b) input–output relationship.

The performance of the half-wave rectifier can benefit from using a fast op. amp. As the input voltage passes through the zero voltage level in a positive direction, the op. amp. output has to quickly jump by about -0.6 V to reverse the previous states of the diodes. During the time it takes to do this the desired relationship $V_o = (-R_2/R_1)V_i$ is not obeyed. This time is minimised by specifying an op. amp. with high slew-rate and bandwidth.

Full-wave precision rectification can be obtained from the circuit in Fig. 9.19. The op. amp. A_1 forms a half-wave rectifier with output V_{hw}. Op. amp. A_2 forms a two-input summing amplifier with unequal input resistors. Thus:

$$V_o = -V_i - 2.V_{hw} \tag{9.21}$$

Consider a negative V_i applied at the input. In this case the half-wave rectifier gives zero output, and substituting $V_{hw} = 0$ into Equation 9.21 results in

$$V_o = -V_i - 2(0) = -V_i \tag{9.22}$$

This relationship gives the left-hand sloping characteristic in Fig. 9.19(b). For a positive applied V_i the half-wave rectifier provides an output which is the inverse of the input: $V_{hw} = -V_i$. On substituting this condition, Equation 9.21 becomes

$$V_o = -V_i - 2(-V_i) = +V_i \tag{9.23}$$

This explains the other sloping region on the right hand side of Fig. 9.19(b). Thus the circuit output is equal to the magnitude of the input voltage and is not affected by the input sign; that is

Because of this relationship the circuit is also called the *absolute value* circuit.

$$V_o = |V_i| \tag{9.24}$$

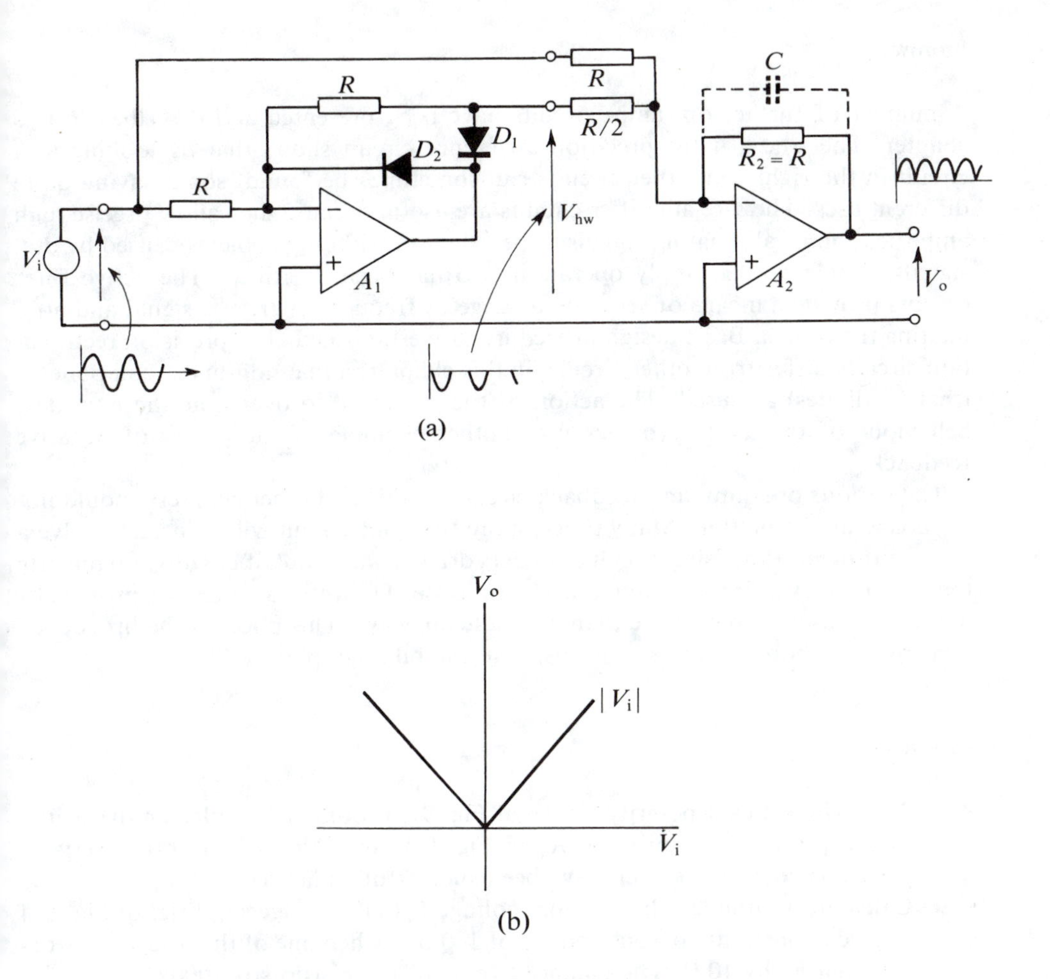

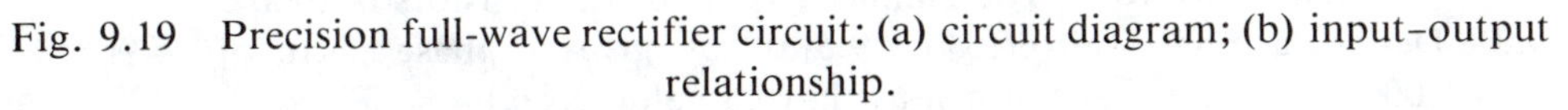

Fig. 9.19 Precision full-wave rectifier circuit: (a) circuit diagram; (b) input–output relationship.

If amplification is required, R_2 must be greater than R. Then

$$V_o = \frac{R_2}{R} |V_i| \tag{9.25}$$

A useful application of the full-wave rectifier circuit is to provide an a.c. measuring function to a d.c. measuring voltmeter. To do this a voltmeter connected to the output needs to be presented with a voltage equal to the average value of the full-wave rectified waveform. This can be achieved by smoothing out the output ripples using the capacitor C shown. The combination of R_2 and C acts as a first-order low-pass filter and to achieve the necessary degree of smoothing C is chosen to give a corner frequency $1/(CR_2)$ which is many times lower than the frequency of the a.c. signal. Greater smoothing can be obtained by interposing a second-order low-pass active filter between the full-wave rectifier and the voltmeter. This filter could be of a type described earlier in this chapter.

Summary

A number of further op. amp. circuits have been presented and described in this chapter. The study of the precision difference circuit shows that by looking at a circuit in the right way other circuits can sometimes be found, some having quite different uses. The a.c. amplifier circuits are useful because they allow precise high amplification of alternating signals to be achieved without problems caused by d.c. signals. Single power supply operation also has been described. The active filter circuits provide a means of selecting a range of frequencies from a signal and attenuating the others. Basic design procedures are established. The precision rectification circuits differ from other circuits in this chapter in that non-linear components (that is, diodes) are used. The action of the op. amp. to overcome the non-ideal behaviour of the diode provides yet another example of the power of negative feedback.

The various op. amp. and feedback circuits in this and other chapters should not be looked at in isolation. Many times, more than one circuit will be used to solve a design problem. Practising circuit designers draw from a pool of circuits. Frequently they will modify a circuit to suit a special purpose. This tutorial guide has provided a modest but useful pool (more than a puddle anyway). This pool can be broadened by consulting books such as those listed in the bibliography.

Problems

9.1 The switched gain-polarity circuit of Fig. 9.2 is obtained by placing the switch in the position occupied by R_4 in Fig. 9.1. In which of the other resistor positions could the switch have been successfully placed?

9.2 Calculate a suitable value of input voltage V_i to the bridge amplifier of Fig. 9.4 to give a maximum output voltage of 100 mV when one of the bridge resistors is increased by 10 Ω. The balanced value of the resistors is 1000 Ω.

9.3 Design a phase-shifter, Fig. 9.5(a), to give a phase shift of $-90°$ at 1000 rad/s. The capacitor must have a value 0.1 μF.

9.4 Design an inverting a.c. amplifier, Fig. 9.8, to have a mid-band gain and input resistance of -100 and 2 kΩ respectively. The lower cut-off frequency is to be at 100 Hz.

9.5 Calculate suitable resistor values for R_1 and R_2 for the bootstrapped amplifier of Fig. 9.8 given that $R_3 = 90$ kΩ, the mid-band gain is to be $+11$ and the circuit is to be bias compensated.

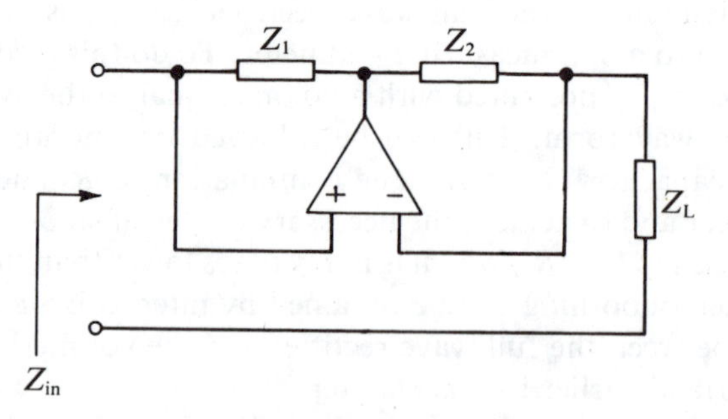

Fig. 9.20 Circuit for Problem 9.7.

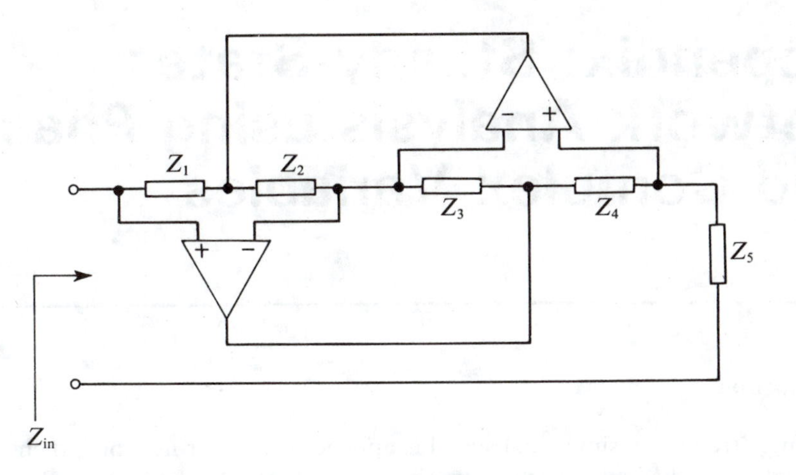

Fig. 9.21 Generalized impedance converter (Problem 9.8).

9.6 Repeat Worked Example 9.3 but with a specified corner frequency at 1000 rad/s.

9.7 For Fig. 9.20 it can be shown that $Z_{in} = -Z_1 Z_L / Z_2$. Using resistors and capacitors only, find ways this circuit can simulate (i) a negative resistor (ii) a negative inductor.

9.8 The circuit in Fig. 9.21 is a generalized impedance converter (g.i.c.). It can be shown that $Z_{in} = +Z_1 Z_3 Z_5 / (Z_2 Z_4)$. Using resistors and capacitors only, find ways to make a simulated positive-value inductor.

9.9 What happens to the input–output relationship of the precision half-wave rectifier circuit of Fig. 9.18 if the directions of both diodes are reversed?

9.10 Repeat Problem 9.9 for the full-wave circuit in Fig. 9.19.

Appendix: Steady-State Network Analysis using Phasors and Complex Variables

Introduction

If a single frequency sinusoidal signal is applied to a linear network, then after any transients have died away the system settles down to a steady-state condition. At all points in the network signals are sinusoidal at the same frequency but have various amplitudes and phase angles.

A wide range of important applications in electrical and electronic engineering contain networks operating under steady-state conditions to sinusoidal excitations.

Any sinusoidal signal has a steady-state waveform of the form

$$v(t) = V \sin(\omega t + \phi) \quad \text{(A.1)}$$

where V is the peak value of $v(t)$, ω is the *angular velocity* and equals the frequency of the signal multiplied by 2π, and the *phase angle*, ϕ, is included in the expression to account for the sinusoidal signal starting at some non-zero amplitude when $t = 0$.

The analysis of the network consists of doing various operations on the signals and component values of the system. Consider for example the operation of adding two sinusoidal quantities,

$$v(t) = v_1(t) + v_2(t)$$

where

$$v_1(t) = V_1 \sin(\omega t + \phi_1) \quad \text{and} \quad v_2(t) = V_2 \sin(\omega t + \phi_2).$$

Thus

$$v(t) = V_1 \sin(\omega t + \phi_1) + V_2 \sin(\omega t + \phi_2) \quad \text{(A.2)}$$

The analytical evaluation of the right-hand side is not too difficult using the appropriate trigonometric expansion formula, however, it is somewhat tedious. Calculations using trigonometric methods become progressively more tedious as the size of the network increases.

The use of the phasor representation of signals and complex-algebra methods allow the steady-state behaviour of linear networks for sinusoidal signals to be carried out in simpler ways. Moreover, they permit circuits containing inductances and capacitances to be transformed so that classical network analysis methods derived originally for resistive circuits can be applied simply to those more general circuits.

For in-depth treatments on this subject consult texts on circuit theory such as: Bobrow, L.S. *Elementary Linear Circuit Analysis* (Holt-Saunders, 1981); Fidler, J.K. *Introductory Circuit Theory* (McGraw-Hill, 1980); Meadows, R.G. *Electrical Network Analysis* (Penguin, 1972).

This appendix provides an outline of these methods.

Phasor Representation

The steady-state sinusoidal waveform $v(t) = V \sin(\omega t + \phi)$ is shown on the right-hand side of Fig. A.1. To the left is a rotating vector representation of $v(t)$. Here a

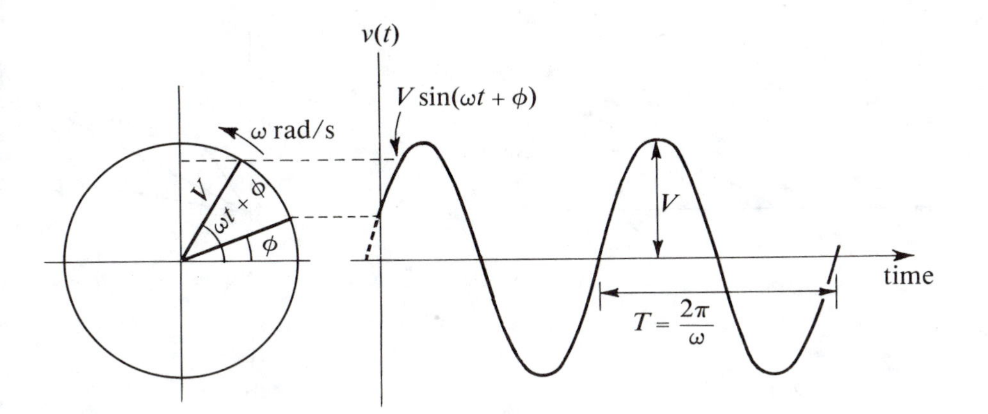

Fig. A.1 Rotating vector representation of sinusoidal waveform.

vector of length V rotates about the origin and its tip traces out a circular path. It moves through a continuously increasing angle $(\omega t + \phi)$ radians starting at ϕ radians at time $t = 0$. The speed of rotation (that is the *angular velocity*) is ω radians per second. It can be seen that the vertical height of the vector, as obtained by projecting it onto the vertical axis, is equal to the waveform $v(t)$. By simple trigonometry this is equal to the vector length multiplied by the sine of the angle through which it has rotated, that is $V \sin(\omega t + \phi)$. Thus the projection of the rotating vector onto the vertical axis generates the sinusoidal waveform $v(t) = V\sin(\omega t + \phi)$.

One period, T, of the sinewave is generated when the angle of the vector moves through one full circle thus $\omega T = 2\pi$ radians. Thus $T = 2\pi/\omega$ as marked.

Networks can be analysed using vector representations of the signals. The need to rotate all the vectors at ω rad/s is an encumbrance which turns out not to be necessary. The point in time $t = 0$ can be chosen as reference and the positions of any rotating vectors noted. A rotating vector now has the stationary vector representation shown in Fig. A.2(a). This is called a *phasor*, and is denoted by $\hat{V}$. The magnitude of the phasor is unaltered at V, and the stationary angle ϕ is the phase constant in the time-domain representation of the signal, $v(t) = V\sin(\omega t + \phi)$. Fig. A.2(a) is referred to as an *Argand diagram*.

Magnitude V corresponds to the peak of the sinewave. Where r.m.s. quantities are desired, the well-known relationship $V_{\text{r.m.s.}} = 1/\sqrt{2}\, V$ is used.

The phasor is defined by its length V and angle ϕ and in *polar form* is written as

$$\hat{V} = V\underline{/\phi} \quad \text{(A.3)}$$

Operations on phasors are usually simpler than on sinewave functions. For example, to add two sinusoidal voltages $v_1(t)$ and $v_2(t)$ of the same angular velocity, as defined by Equation A.2, the phasor diagram in Fig. A.2(b) is constructed. The parallelogram rule can be used to find the resultant phasor $\hat{V}$ as indicated. This resultant phasor is the phasor representation of the sum $v(t) = v_1(t) + v_2(t)$, and so $v(t)$ is generated simply by rotating $\hat{V}$ at ω rad/s and projecting it onto the vertical axis of the Argand diagram, Fig. A.2(b).

In a later section it will be shown that phasors can also be added using complex variables.

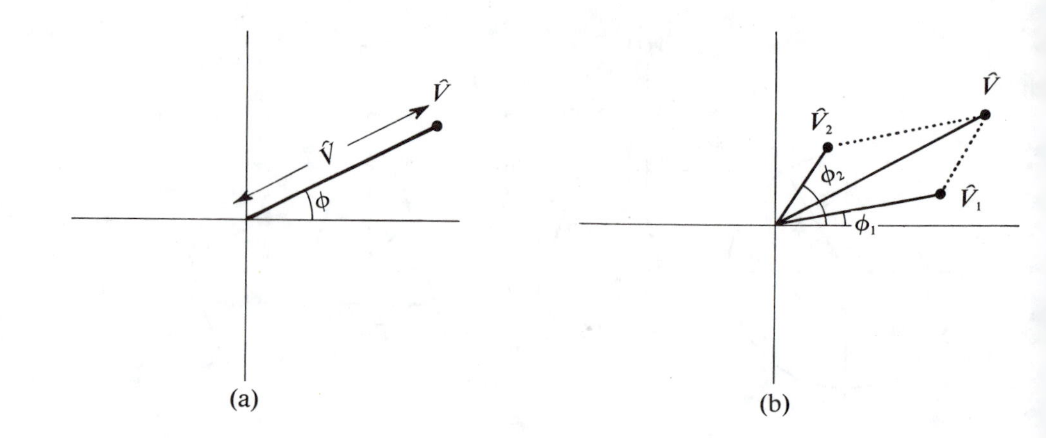

Fig. A.2 Argand diagrams: (a) phasor representation; (b) phasor addition of two voltages.

Complex Representation using the j-Operator

By definition the operator j when applied to a vector $\hat{V}$ causes its phase angle to be increased (that is advanced in the anticlockwise direction) by a quarter of a circle, 90°. Thus

$$\text{if } \hat{V} = V\underline{/\phi} \text{ then } j\hat{V} = V\underline{/\phi + 90°}$$

The operator can be applied a second time

$$j(j\hat{V}) = j^2\hat{V} = V\underline{/\phi + 180°}$$

and a third time,

$$j(j(j\hat{V})) = j^3\hat{V} = V\underline{/\phi + 270°}$$

These operations are shown in Fig. A.3. The double operation j^2 causes a phase of 180° to be added. This is the same as multiplying the vector by -1. Thus $j^2\hat{V} = -\hat{V}$. The triple operation j^3 gives the result $j^3\hat{V} = j^2(jV) = -(jV)$. Applying the operator four times rotates $\hat{V}$ to its original position, thus $j^4\hat{V} = \hat{V}$. This is the same as doing no operation on $\hat{V}$ at all; that is, it is the same as multiplying $\hat{V}$ by unity. Finally, consider the operation j^{-1}. This is the inverse of operator j and causes a backwards rotation of 90°. Reference to Fig. A.3 confirms that this is the same as a forwards rotation due to j^3. Thus $j^{-1}\hat{V} = j^3\hat{V}$, and using the relationship already derived for j^3 we have $j^{-1}\hat{V} = -j\hat{V}$. These properties of the j-operator are now summarised.

$$j^2 \equiv -1;\ j^3 \equiv -j;\ j^4 \equiv 1;\ j^{-1} = \frac{1}{j} \equiv -j \qquad \text{(A.4)}$$

Using these properties any multiple application of the j-operator can be reduced to at most a single j together with a possible sign reversal.

The j-operator can be used to represent phasors. This important step leads to the use of complex algebra methods which provide powerful and convenient ways to

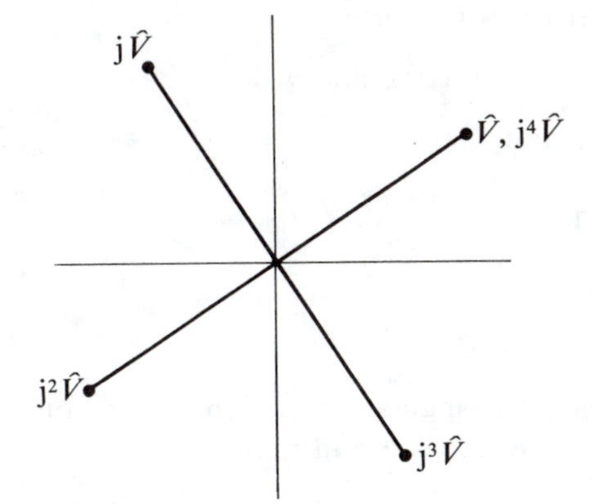

Fig. A.3 Action of the j-operator.

characterize and analyse networks. Consider a phasor $\hat{V} = V\underline{/\phi}$ resolved into its horizontal and vertical components as shown in Fig. A.4. Simple trigonometry gives

$$a = V.\cos\phi \text{ and } b = V.\sin\phi \qquad \text{(A.5)}$$

These two components are themselves phasors; $\hat{a} = a\underline{/0^\circ}$ and $\hat{b} = b\underline{/90^\circ}$. Therefore phasor $\hat{V}$ can be written

$$\hat{V} = \hat{a} + \hat{b} = a\underline{/0^\circ} + b\underline{/90^\circ} \qquad \text{(A.6)}$$

The definition of the j-operator allows the 90° phase shift in the $b\underline{/90^\circ}$ term in the above equation to be replaced by $j(b\underline{/0^\circ})$. Thus $\hat{V} = a\underline{/0^\circ} + j(b\underline{/0^\circ})$. Deleting the $\underline{/0^\circ}$ occurrences gives the *rectangular* form of representation of $\hat{V}$,

rectangular form of phasor: $\hat{V} = a + jb$

Here the real number a is conventionally referred to as the *real part* and b as the *imaginary part*. The whole, $a + jb$, is called a *complex number*.

Rules for converting from one form of phasor to another are readily derived from basic trigonometry applied to Fig. A.4. The rules are

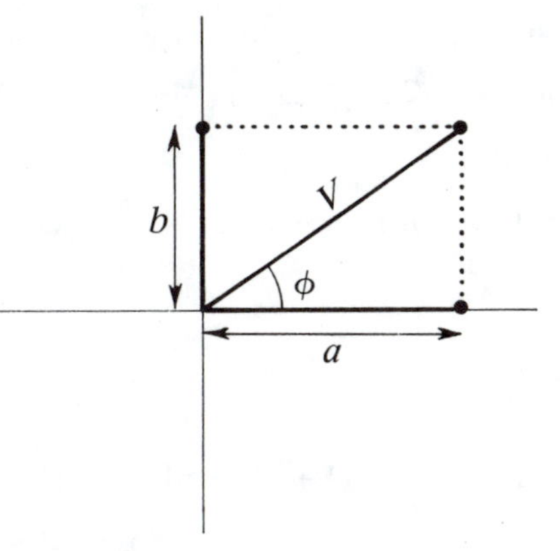

Fig. A.4 Phasor resolved into horizontal and vertical components.

To rectangular form from polar form,

$$\hat{V} = a + jb, \text{ where } a = V.\cos\phi \text{ and } b = V.\sin\phi \tag{A.7}$$

To polar form from rectangular form,

$$\hat{V} = V\underline{/\phi}, \text{ where } V = \sqrt{(a^2 + b^2)} \text{ and } \phi = \tan^{-1}\frac{b}{a} \tag{A.8}$$

Worked Example A.1 Find the polar and rectangular representations of the signal $v(t) = 70\sin(500t + \pi/12)$, where angles are in radians.

Solution: A comparison with Equation A.1 gives $V = 70$, $\omega = 500$ rad/s, $\phi = \pi/12$ rad (that is 15°). The angular velocity ω is not required for the representation of the phasor.

From Equation A.3 the polar form follows directly,

$$\text{polar form: } \hat{V} = V\underline{/\phi} = 70\underline{/15°}$$

The rectangular form is obtained from rule (A.8); $a = V\cos\phi = 70\cos 15° = 67.6$, and $b = V\sin\phi = 70\sin 15° = 18.1$. Thus

$$\text{rectangular form: } \hat{V} = 67.6 + j18.1$$

Operations on Phasors

The phasor representation allows a variety of useful operations to be performed using convenient algebraic methods. These operations are now demonstrated by means of numerical examples.

Two phasors $\hat{V}_1$ and $\hat{V}_2$ will be used whose values in rectangular form are taken to be

$$\hat{V}_1 = 4 + j3 \text{ and } \hat{V}_2 = 1 + j2$$

As an exercise check this polar form is correct.

In polar form these are

$$\hat{V}_1 = 5\underline{/36.9°} \text{ and } \hat{V}_2 = 2.24\underline{/63.4°}$$

Addition

$$\hat{V} = \hat{V}_1 + \hat{V}_2 = (4 + j3) + (1 + j2) = 4 + 1 + j3 + j2$$

Thus

$$\hat{V} = 5 + j5$$

Subtraction

$$\hat{V} = \hat{V}_1 - \hat{V}_2 = (4 + j3) - (1 + j2) = 4 - 1 + j3 - j2$$

Thus

$$\hat{V} = 3 + j1$$

Multiplication

$$\hat{V} = \hat{V}_1 . \hat{V}_2 = (4 + j3) \times (1 + j2)$$
$$= (4 + j3) \times (1) + (4 + j3) \times (j2)$$
$$= 4 + j3 + j8 + j^2 6 = 4 + j11 + j^2 6$$

But by properties A.4, $j^2 \equiv -1$, hence

$$\hat{V} = 4 + j11 + (-1)6 = -2 + j11$$

Alternatively, the polar form can be employed for multiplication by using the properties that moduli multiply and angles add:

$$\hat{V} = \hat{V}_1 \times \hat{V}_2 = (5\underline{/36.9^\circ}) \times (2.24\underline{/63.4^\circ})$$
$$= (5 \times 2.24)\underline{/36.9^\circ + 63.4^\circ} = 11.2\underline{/100.3^\circ}$$

The reader may confirm that converting this to rectangular form gives the same value, $\hat{V} = -2 + j11$, as obtained by multiplying the phasors using the rectangular forms of phasor representation.

Division

$$\hat{V} = \frac{\hat{V}_1}{\hat{V}_2} = \frac{4 + j3}{1 + j2}$$

The *complex conjugate* of a complex number will be used to perform this calculation. This is obtained by reversing the sign of the imaginary part of the complex number. An asterisk (*) is usually used to denote the operation of taking the complex conjugate, thus for $\hat{V}_2 = 1 + j2$ then $\hat{V}_2^* = 1 - j2$. Division by a complex number is aided by multiplying the numerator and denominator by the complex conjugate of the denominator. Thus

$$\hat{V} = \frac{4 + j3}{1 + j2} = \frac{(4 + j3)(1 - j2)}{(1 + j2)(1 - j2)} = \frac{(4 + j3)(1) + (4 + j3)(-j2)}{(1 + j2)(1) + (1 + j2)(-j2)}$$
$$= \frac{4 + j3 - j8 - j^2 6}{1 + j2 - j2 - j^2 4} = \frac{4 - j5 - j^2 6}{1 - j^2 4} = \frac{4 - j5 - (-1)6}{1 - (-1)4} = \frac{10 - j5}{5}$$

Multiplying numerator and denominator by the same amount is allowed because it does not alter the value of the expression.

Hence $\hat{V} = 2 - j1$.

The trick used here of multiplying by the complex conjugate converts the denominator into a real number which is easily divided into the numerator to complete the calculation. This trick works for any denominator complex number.

The polar representation can be used to perform phasor division using the properties that moduli divide and angles subtract. The calculation is as follows:

$$\hat{V} = \frac{\hat{V}_1}{\hat{V}_2} = \frac{5\underline{/36.0^\circ}}{2.24\underline{/63.4^\circ}} = \frac{5}{2.24}\underline{/36.9^\circ - 63.4^\circ}$$

Thus $\hat{V} = 2.24\underline{/-26.5^\circ}$.

The reader again may confirm that converting this to rectangular form gives the same value, $\hat{V} = 2 - j1$ as obtained by using the rectangular form of phasor representation to perform the division.

It will be observed that both multiplication and division are easier to perform in polar form. This assumes that the polar forms are already available. If a conversion from rectangular form to polar form is required to be carried out first, then a saving of effort is not necessarily obtained.

Phasor Analysis of Circuits

Consider the simple series RLC circuit in Fig. A.5(a) which is excited by a sinewave current source $i(t) = I\sin(\omega t + \phi)$. The voltage drop across the circuit $v(t)$ is given by the sum of the individual voltage drops across R, L and C. Assuming L and C carry no initial current or voltage,

Note $\cos(\theta) = \sin(\theta + 90°)$.

$$
\begin{aligned}
v(t) &= R.i(t) + L\frac{\mathrm{d}i(t)}{\mathrm{d}t} + \frac{1}{C}\int i(t).\mathrm{d}t \\
&= R.I\sin(\omega t + \phi) + L.\frac{\mathrm{d}}{\mathrm{d}t}\{I\sin(\omega t + \phi)\} + \frac{1}{C}\int I\sin(\omega t + \phi)\mathrm{d}t \\
&= R.I\sin(\omega t + \phi) + LI\omega\cos(\omega t + \phi) + \frac{1}{C}I\left(\frac{-1}{\omega}\right)\cos(\omega t + \phi) \\
&= R.I\sin(\omega t + \phi) + \omega LI\sin(\omega t + \phi + 90°) - \frac{1}{\omega C}I\sin(\omega t + \phi + 90°)
\end{aligned}
$$

Because the right-hand sides consist of sinusoids of the same angular velocity ω, the equations can be represented in phasor form. The second and third sine terms contain additional 90° phase constants so each will attract a j operator. The resulting phasor equation is

$$\hat{V} = R.\hat{I} + \mathrm{j}\omega L\hat{I} - \frac{\mathrm{j}}{\omega C}\hat{I},$$

where $\hat{I} = I\underline{/\phi}$

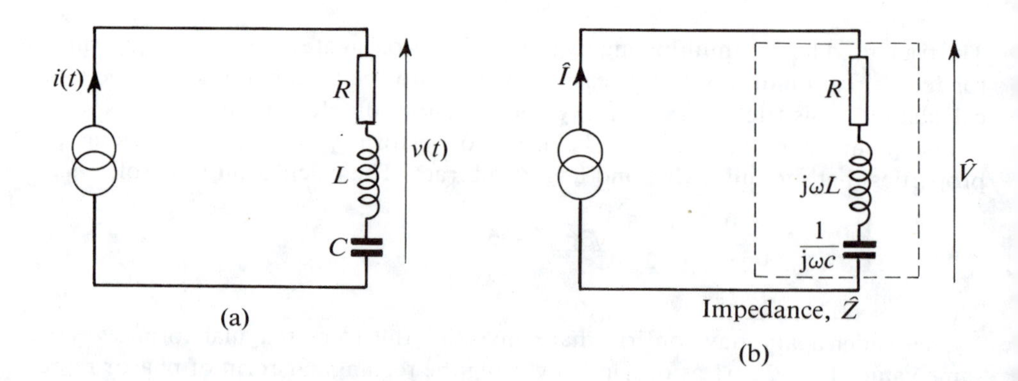

Fig. A.5 Simple RLC circuit: (a) circuit in the time-domain; (b) phasor transformation of the circuit.

Factorising this gives

$$\hat{V} = \left(R + \mathrm{j}\omega L + \frac{1}{\mathrm{j}\omega C}\right) \hat{I} \tag{A.9}$$

Note from Equation A.4 that $-\mathrm{j} = 1/\mathrm{j}$.

Given values for the various parameters, this expression can be readily evaluated using the methods for performing operations on phasors which are given in the previous section.

Equation A.9 reveals a very important and far-reaching property. The equation expresses an Ohm's law type of relationship in which the overall ohmic quantity in parentheses contains the resistance R and two other ohmic terms ($\mathrm{j}\omega L$) and ($1/\mathrm{j}\omega C$) each due to the inductor and capacitance. These terms are called *reactances*. The expression A.9 can be obtained directly from a simple transformation of the circuit. All components are replaced by their phasor equivalents. The result for this circuit is shown in Fig. A.5(b). The ohmic quantities resistance R, inductive reactance $\mathrm{j}\omega L$, and capacitive reactance $1/\mathrm{j}\omega C$ are all in series and by inspection have a combined *impedance* given by

$$\hat{Z} = R + \mathrm{j}\omega L + \frac{1}{\mathrm{j}\omega C}$$

By Ohm's law $\hat{V} = \hat{Z}.\hat{I}$ and on substituting for $\hat{Z}$ the Equation A.9, previously obtained rather laboriously, now simply follows.

The circuit transformation, as illustrated by this example, can be applied generally to any electrical network. The transformed circuits can be analysed using classical techniques for d.c. resistance circuits based on Ohm's law, Kirchoff's laws etc. This provides a powerful and effective means of analysing electrical networks under sinusoidal steady-state conditions.

Beyond the j-operator: the Laplace Transform

Though the phasor and j-operator methods are very useful for analysing the steady-state analysis of networks, they are unable to help where the analysis of transient effects in the networks is required. For example, if $i(t)$ in A.5(a) is a pulse source of short duration the analysis to find $v(t)$ becomes a transient analysis problem. Because $i(t)$ is not a continuous sinusoid the phasor representation and analysis methods cannot be applied.

Laplace transform methods allow transient problems to be solved. A proper discussion of these methods is outside the scope of this tutorial guide. However, one thing can be mentioned. That is that the Laplace transform, which is a special integral, transforms the signals and circuit components into functions of a variable s called the *complex-frequency* variable or the *Laplace* variable. This variable is a complex quantity and is given by $s = \sigma + \mathrm{j}\omega$. Here $\mathrm{j}\omega$ is the same $\mathrm{j}\omega$ that appears in the reactance terms $\mathrm{j}\omega L$ and $1/\mathrm{j}\omega C$. Thus the Laplace transform of the circuit A.5(a) is identical to that in Fig. A.5(b) with $\mathrm{j}\omega L$ and $1/\mathrm{j}\omega C$ replaced by sL and $1/sC$ respectively. Also the currents and voltages are now Laplace transform functions. Steady-state sinusoidal analysis is obtained simply from the Laplace transform analysis by setting s equal to $\mathrm{j}\omega$ and interpreting the currents and voltages as phasors.

Books which explain Laplace transforms include Meade, M.L., and Dillon, C.R. *Signals and Systems* (Van Nostrand, 1986) and Stroud, K.A. *Laplace Transforms: Programmes and Problems* (Stanley Thornes, 1985).

Laplace transform techniques are not required for this tutorial guide.

Bibliography

The following books provide good sources for more information on feedback amplifiers and op. amp. circuits and applications.

Bowron, P. and Stephenson, F.W. (1979) *Active Filters for Communications and Instrumentation*, McGraw-Hill.

Clayton, G.B. (1975) *Linear Integrated Circuit Applications*, Macmillan.

Clayton, G.B. (1979) *Operational Amplifiers*, 2nd edn, Butterworths.

Graeme, J.G. (1973) *Applications of Operational Amplifiers: Third-Generation Techniques*, McGraw-Hill.

Graeme, J.G. (1977) *Designing with Operational Amplifiers: Application Alternatives*, McGraw-Hill.

Horowitz, P. and Hill, W. (1980) *The Art of Electronics*, Cambridge University Press.

Jung, W.G. (1974) *IC Op-Amp Cookbook*, Howard W. Sams.

Marston, R.M. (1975) *110 Operational Amplifier Projects for the Home Constructor*, Newnes-Butterworths.

Rosenstark, S. (1986) *Feedback Amplifier Principles*, Macmillan.

Rutkowski, G.B. (1984) *Handbook of Integrated-Circuit Operational Amplifiers*, 2nd edn, Prentice-Hall.

Sheingold, D.H. (ed.) (1976) *Nonlinear Circuits Handbook*, Analog Devices, Inc.

Tobey, G.E., Graeme, J.G. and Huelsman, L.P. (eds) (1971) *Operational Amplifiers: Design and Applications*, McGraw-Hill.

Wait, J.V., Huelsman, L.P. and Korn, G.A. (1975) *Introduction to Operational Amplifier Theory and Applications*, McGraw-Hill.

Answers to Numerical Problems

2.1 (i) PFB, (ii) PFB, (iii) PFB, (iv) NFB, (v) NFB
2.2 Cases (i) and (ii) are likely to be unstable.
2.3 $\beta = -0.098$
2.4 Intended closed-loop gain is 75.
2.5 Actual value of A was $-99\,900$
2.6 Sensitivity is 0.0476
2.7 $A = 4000$
2.8 Noise output $= 0.476 \times 10^{-3}$ V

3.1 48 mA and 48 V
3.3 For c.c.c.s. representation $K_I = 200$; for v.c.c.s. representation $K_G = 0.02$ S; for c.c.v.s. representation $K_R = 1$ MΩ. In all representations $r_{in} = 10$ kΩ and $r_{out} = 5$ kΩ.
3.4 If $r_{in} = 0$ no input voltage can be developed at the input. Hence only the current controlled representations of Figs. 3.4a and b exist.
3.5 $V_{out} = -4.926$ V, $i_{out} = -0.4926$ mA, $A_V = -2.956$, $A_I = -29.560$.
3.6 $V_{out} = -4.954$ V, $i_{out} = -0.4954$ mA, $A_V = -2.725$, A_I -54.490. Behaviour altered because coupling factors differ from those in Problem 3.5.
3.7 Minimum $C_1 = 796$ nF, and maximum $C_2 = 19.9$ pF.
3.8 (i) $89.4 \angle -153.44°$, (ii) $44.7 \angle -116.56°$
3.9 $\omega_{HP} = 0.6436\ \omega_2$. At ω_{HP} the gain phase angle is $-65.53°$.
3.10 (i) 10^7 Hz, (ii) intersects at infinite frequency

4.1 -9.08 kΩ, 9.08 Ω, 9.08 Ω respectively.
4.2 486 kΩ, 20.6 Ω, 20.6 Ω respectively
4.3 $R = 995$ Ω, $r_{in} = 300$ kΩ
4.4 13.28, 16.59 Ω, 301.3 kΩ respectively
4.5 $R_1 = 51.63$ kΩ. Also closed-loop gain $= 50$, $r_{if} = 200$ kΩ and $r_{of} = 50$ Ω.
4.6 $A_{If} = 11.00$, $r_{if} = 4.82$ Ω, $r_{of} = 2.08$ MΩ
4.7 $Z_{if} = 1.96 - j\,9.80$ Ω
4.9 Use configuration Fig. 4.5a with forward amplifier being the differential amplifier. After feedback $A_{Vf} = A_{dm}/(1 + A_{dm\beta})$.

5.1 $R_T = -22.22$ kΩ, which leads to $R_{Tf} = -0.956$ kΩ, $r_{if} = 19.14$ Ω, $r_{of} = 21.53$ Ω
5.2 $A_{Vf} = 93.4$, which leads to $A_{Vf} = 5.64$, $r_{if} = 60.8$ kΩ, $r_{of} = 103$ Ω
5.3 $r_{if} = 502$ kΩ, $r_{of} = 502$ kΩ, $G_{Tf} = -0.996 \times 10^{-3}$ S
5.4 $A_I' = 33\,115.8$ which leads to $r_{if} = 15.27$ Ω, $r_{of} = 3.322$ MΩ, and $A_{If} = 100.69$
5.5 $r_{if} = 446.4$ kΩ, $r_{of} = 335.3$ kΩ, $G_{Tf} = -0.996 \times 10^{-3}$ S
5.6 $r_{if} = 11.00$ MΩ, $r_{of} = 3.620$ Ω, $A_{Vf} = 5.994$
5.7 $f_U = 200$ kHz, $f_L = 10$ Hz

5.8 $\omega_0 = \omega_L/\sqrt{3}$
5.9 $\omega = \omega_1\sqrt{[1 + (A\beta/2)]}$
5.10 $\omega = \omega_1/\sqrt{[1 + (A\beta/2)]}$

6.1 (i) $R_1 = 10$ kΩ, $R_2 = 1$ MΩ, (ii) $-15 \leqslant A_V \leqslant -1010$, r_i is constant at 10 kΩ
6.2 (i) $|A_V| = 20$, (ii) $|A_V| = 0.795$
6.3 Put the capacitor in series with R_1 to make a high-pass filter
6.4 $A_V = -99.899$, $Z_i = 100\ 100$ Ω, $Z_0 = 0.11087$ Ω
6.6 $C = 1$ μF, $R = 100$ kΩ
6.7 The output waveform is a linear negative going ramp, which takes 20 s to reach -10 V.
6.8 (i) 26, (ii) 10.965, (iii) 6.894
6.10 (i) V^+ supplies 1 mA, V^- supplies -21 mA, (ii) 264 mW, (iii) 100 mW, (iv) 64 mW

7.1 (i) 0.101 V, (ii) 0.111 V, (iii) 0.103 V
7.2 0.2 μA
7.3 36 mV
7.4 833 s for worst case
7.5 3333 s for worst case
7.6 10 kHz
7.7 10 kHz
7.8 50 kHz
7.9 0.884 V

8.1 $A_{dm} = 3$
8.2 R_A
8.3 $R_B = 90$ kΩ, $R_2 = 100$ kΩ
8.4 $A_{dm} = 99.995$, $A_{cm} = 9.9 \times 10^{-3}$, CMRR = 10 090 = 80.1 dB
8.6 $R = 7.96$ kΩ

9.1 Only in place of R_3. This gives $A_v = -1$ for S open and $A_v = +1$ for S closed.
9.2 $V_i = 200$ V. Relative resistance change $\delta = 10\ \Omega/1000\ \Omega$. Then solve any of Equations 9.7, given $V_i = 100$ mV.
9.3 $R_3 = 10$ kΩ by solving Equation 9.9. Then choose $R_1 = R_2 = R$ where R is any convenient value. Choosing $R = 20$ kΩ gives bias current compensation.
9.4 $Z_{in} = R_1$ at mid-band frequencies, so $R_1 = 2$ kΩ. Thus $R_2 = 200$ kΩ to give $A_v = -R_2/R_1 = -100$. Capacitor C must satisfy $\omega_C = 1/CR_1$. Hence $C = 0.75$ μF.
9.5 Resistors R_1, R_2 must satisfy $A_v = 11 = 1 + R_2/R_1$, and also $R_1 + R_3 = R_2$ for bias current compensation. Hence $R_1 = 10$ kΩ, $R_2 = 100$ kΩ.
9.6 $R_3 = R_4 = 100$ kΩ, $R_1 = 50$ kΩ, $C_5 = 3.54$ nF, $C_2 = 28.3$ nF.
9.7 (i) (R_1, R_2, R_L) gives $Z_{in} = -(R_1R_L/R_2)$, a negative constant resistance. (ii) (R_1, C_2, R_L) gives $Z_{in} = j\omega L$ where $L = (-R_1R_LC_2)$.
9.8 (R_1, C_2, R_3, R_4, R_5) gives $Z_{in} = j\omega L$ where $L = (+R_1R_3R_5C_2/R_4)$. Also (R_1, R_2, R_3, C_4, R_5) gives $L = (+R_1R_3R_5C_4/R_2)$.
9.9 For positive V_i, output V_o is zero. For negative V_i output $V_o = -(R_2/R_1)V_i$, a positive quantity.
9.10 Fig. 9.19b turns upside down. Thus $V_o = -|V_i|$.

Index

A.C. amplifiers 156–61
Active filters 161–70
Amplifier models 21
Amplifiers without feedback 21–46
Amplitude control of oscillator 141
Analogue computation 135
Argand diagram 177

Band-pass response 161
Bandwidth 31
Bias current compensation 121
Bode diagram 32
Bootstrapping 159
Bridge amplifier 152
Butterworth response 165, 169

Closed-loop gain 5
Common-mode amplification 43
Common Mode Rejection Ratio 43
Common-mode voltage 43
Comparator 5
Complex number 179
Conditional stability 94
Corner frequencies 30
Current-controlled current source 22
Current-controlled voltage source 24
Current coupling factor 28
Current ratio 31

D.C. blocking capacitor 157
Decibel 31
Differential amplifier 40
Differential-mode amplification 40
Differential-mode voltage 40
Differentiator circuit 108
Direct coupled amplifier 35
Distortion reduction 13
Dual-in-line encapsulation 97

Error signal 1
External offset balancing 119

Feedback amplifier circuits 47–68
Feedback configurations 47
Feedback fraction 5
First order active filters 163–4
Forward-path gain 4
Frequency compensation 95
Frequency response effects 29
Frequency response of feedback amplifiers 88
Full-power bandwidth 126
Fundamental feedback relationship 6

Gain-bandwidth product 89
Generalized impedance convertor 175

Half-power bandwidth 31
Half-power frequencies 30
High-frequency effects 29, 35
High-pass response 161

Impedance 183
Inverting voltage amplifier 100
Ideal amplifiers
  current amplifier 22
  trans-conductance amplifier 24
  trans-resistance amplifier 24
  voltage amplifier 24
Input/output impedance effect of feedback 49
Input resistance 24
Instability 9, 90
Integrator circuit 107
Integrator drift 123
Instrumentation amplifier 42, 131
Inverse function principle 142

j-Operator 178

Laplace transform 183
Large applied negative feedback 56
LC filters 162
Loading effects 69, 82
Loop gain 6
Low-frequency effects 29

Low-pass response 161

Mid-band gain 29
Mid-band region 29, 32
Modified forward amplifier 71
Multi-loop feedback active filter 167
Multiple-loop feedback 15

Negative feedback definition 7
Noise reduction 13
Non-inverting voltage amplifier 109
Non-linear functional blocks 135
Nyquist diagram 38

Offset null 98
Offset-null, potentiometer 119
Open-loop gain 6
Operational amplifier
  antilog circuit 144
  bias currents 120
  definition 97
  divider circuit 144
  frequency compensation 126
  frequency response 124
  input/output characteristics 99
  internal sections 99
  log circuit 144
  non-idealities 115–30
  offset currents 120, 122
  offset voltages 115
  offset-voltage model 116
  precision difference circuit 131
  triangle wave/square wave generator 145
Operations on phasors 180
Output resistance 24

Pass-band 161
Peaking of response 90
Phase shifter 154
Phaser analysis of circuits 182
Phasor representation 176
Pin-compatibility 97
Positive feedback definition 7
Power supplies to op. amps. 111
Precision difference amplifier 131
Precision rectifier 170–3

Q-factor 165

Rauch structure 167
RC-active filter 139
Real amplifier models 25

Second order active filters 164–70
  selectivity 165
Sensitivity 11
Series-current feedback 49, 61
Series-voltage feedback 49, 59
Short-circuit protection 112
Shunt current feedback 49, 64
Shunt voltage feedback 49, 56
Signal-to-noise ratio 14
Single-ended amplifiers 40
Single power supply operation 159
Slew rate 126
Stop-band 162
Strain gauge transducer 152
Successive integration method 136
Summing amplifier 105
Summing integrator 135
Switched gain-polarity circuit 150

Temperature regulation 1
Tow-Thomas biquad 170
Transfer function realization 138
Two-stage amplifier example 27

Unity buffer 110

Variable bipolar-gain circuit 151
Virtual earth 101
Voltage-controlled current source 24
Voltage-controlled voltage source 24
Voltage coupling factor 27
Voltage ratio 31

Wein bridge oscillator 139